HIROSHIMA

A Study in Science, Politics, and the Ethics of War

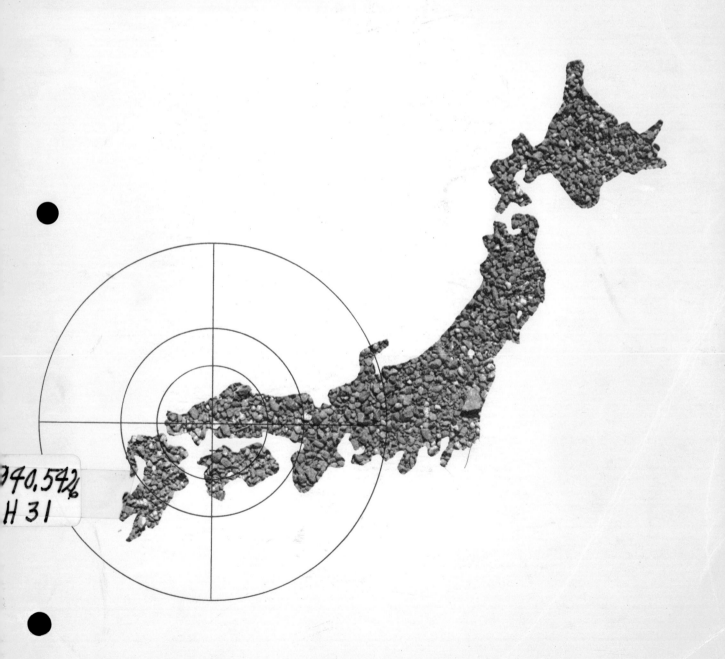

HIROSHIMA

A Study in Science, Politics, and the Ethics of War

JONATHAN HARRIS
Paul D. Schreiber High School
Port Washington, New York

General Editors
RICHARD H. BROWN
VAN R. HALSEY

Developed by The Amherst Project
Committee on the Study of History

ADDISON-WESLEY PUBLISHING COMPANY
MENLO PARK, CALIFORNIA · READING, MASSACHUSETTS
LONDON · DON MILLS, ONTARIO

KOREA AND THE LIMITS OF LIMITED WAR

WHAT HAPPENED ON LEXINGTON GREEN? An Inquiry Into the Nature and Methods of History

FREEDOM AND AUTHORITY IN PURITAN NEW ENGLAND

COMMUNISM IN AMERICA: Liberty and Security in Conflict

HIROSHIMA: A Study in Science, Politics, and the Ethics of War

COLLECTIVE SECURITY IN THE 1930's: The Failure of Men or the Failure of a Principle?

LIBERTY AND LAW: The Nature of Individual Rights

THE EMBARGO OF 1807: A Study in Policy-Making

GOD AND THE GOVERNMENT: Problems of Church and State

CONSCIENCE AND THE LAW: The Uses and Limits of Civil Disobedience

THE AMERICAN WEST AS MYTH AND REALITY

ABRAHAM LINCOLN AND EMANCIPATION: A Man's Dialogue With His Times

IMPERIALISM AND THE DILEMMA OF POWER

Copyright © 1970 by Addison-Wesley Publishing Company, Inc.
Philippines Copyright 1970.

All rights reserved. No part of this publication may be reproduced, stored in a retrieval system, or transmitted, in any form or by any means, electronic, mechanical, photocopying, recording, or otherwise, without the prior written permission of the publisher. Printed in the United States of America. Published simultaneously in Canada.

ISBN: 0-201-02687-2
EFGHIJK-BA-787654

OF HISTORY AND MEANING: A Prologue

All we have of the past is records strewn hither and yon, like bare bones of a skeleton on the desert. When we want to know what happened we go to the bones with questions that grow out of our own experience. It is the questions we ask that enable us to see certain relationships among the bones and, if we fail to ask the right questions, to miss other relationships that were there. In short, it is the questions that give meaning to the data, and these come from us, out of our own world, and reflect the things we want to know. Thus, while the bones are out of the dead past, their meaning or what we call history is very much a part of the live present. It is something we do to the past as we try to better our understanding of ourselves and of the world around us, and to grow as human beings.

So it is, in fact, with the way we confront and learn from nearly every kind of experience, past or present, in or out of school, whether the data we deal with is pot or sex, politics or history. The meaning that adheres to the experience or situation comes from us, from the curiosity or questions we bring to the experience, from our own past experience as it sensitizes us and enables us to *hear* or *see* or *feel* certain things in the situation. A theoretical physicist will see and understand things in a scholarly paper about nuclear weapons which the rest of us can't comprehend. A devotee of rock and roll will hear a piece of jazz music differently from someone who regularly listens to Beethoven. Each of us will *feel* and react differently in the face of a common experience. Learning is an act that each individual does for himself, even when he is learning from and with others. This does not mean that each man's learning is as good as another, for not only do our experiences differ, but our skills as well. It does mean that each man's learning is his own, ultimately a personal matter.

The materials that follow take a new approach to the study of history, based upon these ideas. They do not tell a story or provide a set of answers. They are some bare bones, pieces of evidence from the past, each of them identified. They are presented with the conviction that the study of history is the pursuit of genuine and contemporary questions relevant to the pursuer into evidence from the past. The trail of these questions may lead in various directions. There is no reason why the evidence should be considered in the order in which it is presented here. Nor will all the evidence necessarily prove relevant or responsive to questions that will doubtless vary in meaning from individual to individual and class to class. In this case, particular items of evidence should most logically be omitted or left to some later time when they may seem relevant.

Finally, and most importantly of all, the evidence presented here is only some evidence, which an individual or group of individuals has selected and edited, basically in response to his or their questions. There are more bones where these came from that may and can be added. Those presented here might take on different meaning if they were edited differently. The logic of learning dictates that ultimately each pursuer, each learner, must be free to go behind the confines of the evidence presented here to additional evidence or, if he distrusts the evidence as presented, to go to the original and reconstruct it for himself. For if, as learners, we are pursuing genuine questions that have to do with our own growth and with human understanding, they must necessarily carry us beyond the materials that anyone else can give us, beyond the doors of the classroom and the academy, and ultimately to life itself.

Richard H. Brown

Contents

INTRODUCTION 1

SECTION 1 The Apparent Choice: Japanese Lives vs. American Lives . . 2
    A. The Price Hiroshima Paid 2
    B. The Reasons Why 8

SECTION 2 Was It a Military Decision? 12
    A. Setting the Stage for the Great Decision................. 12
    B. The New Commander-In-Chief 18
    C. Advice from the Military 20
    D. Was There a Decision at All?............................ 23
    E. Evidence after the Fact 25

SECTION 3 The Agony of the Atomic Scientists 31

**SECTION 4 Was It the Product of a Diplomatic Blunder—
or of Diplomatic Calculation?** 40
    A. The "Unconditional Surrender" Problem 40
    B. The Russian Problem 52

SECTION 5 Was It a Morally Defensible Act? 60
    A. Public Conscience in a Decade of War 60
    B. The Conscience of Science 72
    C. The Conscience of Soldiers 74

SUGGESTIONS FOR FURTHER READING 77

INTRODUCTION

Americans tend to react to the awesome fact of the dropping of the atomic bomb on Hiroshima in either of two ways. Some are so revolted that they automatically condemn all who had anything to do with perpetrating such horror. Others accept it as an act of war, brutal perhaps, but necessary.

Ultimately the decision to drop the bomb demands more than judgment on the part of those who look back. It was a complex decision, involving not only military considerations but politics and diplomacy. It raised the question, in a single tragic flash, of the role of science in the lives of modern man. Above all, it raised ethical questions that are with us still today, and to which there are no easy answers. In short, it was human drama of such magnitude and complexity that to try to understand it is to try to understand ourselves.

This unit goes back to that climactic time in World War II when the decision to drop the bomb was still in the making. It has to do with the fantastic complexity of the problems that pressed down upon the men who bore the responsibility of decision. It may seem easy and tempting to suggest answers to their problems from the comfortable vantage point of the present. It will be less so to the degree that we make the effort to place ourselves within the complicated and chaotic context of those war years, when these problems were real.

SECTION 1

The Apparent Choice:
Japanese Lives vs. American Lives

It is difficult for some Americans to consider the bombing of Hiroshima dispassionately, because of the overwhelming loss of human life involved. It may, therefore, be helpful to consider these emotion-charged aspects at the beginning in order to try to decide how much importance they deserve in the formation of any final conclusion.

This section consists of two contrasting parts. Part A tells what happened to some of the people of Hiroshima. Part B focusses on the other side of the story, indicating the reasoning behind the American decision to use the atomic bomb.

A. THE PRICE HIROSHIMA PAID

The atomic bomb affected the lives of its victims in many different ways. The documents in Part A describe first its immediate impact as experienced by some of its younger victims; then its medical effects as reported by scientists; and finally its lingering aftereffects twenty years later.

1. In 1951 Dr. Arata Osada, a distinguished Japanese educator, asked the young people of Hiroshima to write their personal memories of that day in 1945, six years earlier, when the atomic bomb destroyed their city. A brief sampling of portions of their vivid recollections follows:[1]

—5th grade girl. 4 years old in 1945.—

> We were just about to eat breakfast—we were just about to put our chopsticks into our mouths.... Just as we saw a bright flash there was a loud bang and I almost fainted.... Mother, while she was trying to rescue a child who lived next door, touched poison[2]....
>
> Since Mother was in great pain day after day, we called the doctor. The doctor said, "The baby is going to be born pretty soon." At the end of August a baby was born. But only the baby's head was born and then the baby and Mother died together. I was terribly sad....

---

[1] *Children of the A-Bomb: Testament of the Boys and Girls of Hiroshima,* compiled by Dr. Arata Osada (original Japanese edition, Tokyo, Iwanami Shoten, 1959; first American edition, New York, Putnam, 1963), pp. 13, 14-15, 79-81, 83, 146-147, 165, 234-239 *passim.*

[2] The expressions "breathed poison," "touched poison," and "breathed gas" were in common use at the time to explain symptoms that were later understood to be caused by radio-active substances released by the explosion, i.e., "radiation sickness."

—11th grade boy. In 5th grade in 1945.—

I saw several people plunging their heads into a half-broken water tank and drinking the water. I was very thirsty too, and I was so happy to see some people again that without thinking I left my parents' side and went toward them. When I was close enough to see inside the tank I said "Oh!" out loud and instinctively drew back. What I had seen in the tank were the faces of monsters reflected from the water dyed red with blood. They had clung to the side of the tank and plunged their heads in to drink and there in that position they had died. From their burned and tattered middy blouses I could tell that they were high school girls, but there was not a hair left on their heads; the broken skin of their burned faces was stained bright red with blood. I could hardly believe that these were human faces. . . .

—7th grade girl. In 1st grade in 1945.—

[We] were eating breakfast at the time. From no particular place there was a bright flash.

Surprised, I asked, "Mother, what was that!"

And just as she answered, "What could it be!" we heard a tremendous explosion and everything became pitch dark and I couldn't even see the face of my mother who was right beside me.

In that second some heavy object fell and pressed me down painfully.

When I frantically yelled, "Mother, Mother!" there was a voice calling, "Set-chan, Set-chan." Mother's voice came through broken and weak.

I shouted, "Setsuko is here, Setsuko is here!"

And then I heard a voice saying, "Set-chan, try to move your legs."

So I kicked as hard as I could. Just at my chest there was a scraping sound of something being moved. After a little while I was finally able to crawl out of there.

"Mommy, hurry and get up," I said and pulled her hand; but she said, "It hurts, it hurts," and she didn't try to move.

Frightened, I looked carefully and saw that there was a big beam lying across her right arm and her back. No matter how hard I tried to move it my strength wasn't enough to even budge it. With her left hand which she was able to move Mother had helped me out. I took hold of that heavy beam and tried desperately to move it but it had not the slightest effect.

From all sides I could hear voices calling, "Help, help!"

At the top of my voice I yelled, "HELLLLLP!" but there wasn't a single person to come.

The fire had already come close to me. Everytime the flames licked out, my hair got singed.

Desperately I said, "Mommy, Mommy—hurry!" but nothing could be done.

Mother was saying urgently, "Mother will come after you, Set-chan, so you get away first. Now quickly, quickly."

I was determined not to escape without my mother. But the flames were steadily spreading and my clothes were already on fire and I couldn't stand it any longer. So screaming, "Mommy, Mommy!" I ran wildly into the middle of the flames. No matter how far I went it was a sea of fire all around and there was no way to escape. So beside myself I jumped into our water tank. The sparks were falling everywhere so I put a piece of tin over my head to keep out the fire. The water in the tank was hot like a bath. Beside me there were four or five other people who were all calling someone's name. While I was in the water tank everything became like a dream and sometime or other I became unconscious.

I don't know how many hours passed but when I regained consciousness it seemed to be morning. There was still smoke rising here and there and sizzling noises of things burning. Right beside me I found a woman still in the water who seemed to be asleep, but she was dead.

I suddenly became frightened and I called in a loud voice to a man who was passing, "Please come and help me."

The man came and lifted me out of the tank and he told me that if I go straight ahead there is a medical relief station there; so I walked there with him.

While I was thinking, 'I wonder what has happened to Mother,' I looked at my own arms and saw that the skin was all burned and broken up with wounds.

My chest and my back began to hurt more and more, so when we arrived at the relief station they fixed me up. . . .

On a certain day five days after that Grandmother said, "Set-chan's Mother has indeed died after all."

We knew because one of my uncles came back bringing her ashes. Mother had finally died just as I had left her. Holding the urn in my arms I lifted up my voice and wailed. The tears kept on and on. The two cousins and my aunt who had been with her then had all died too.

It was decided that I would be brought up in my grandmother's house. From about then my hair gradually fell out and the burns on my arms got worse. . . .

When I was a fourth grader I got tuberculosis and I spent five months in the Japan Red Cross Hospital. They said this was also the result of the A-bomb. When I left the hospital I was so thin I was just like a skeleton. . . .

—*11th grade boy. In 5th grade in 1945.*—

On the morning of the 6th, my mother, who was to have gone back to Ogata, was in the vestibule paying the bill, and she was talking to the proprietor of the inn. My baby brother, who was only a year and two months old, and the man from the lighthouse were there with her. I was lying on the sofa in the lounge having a lot of fun teasing the cat. Suddenly from outside the front entrance an indescribable color and light—an eerie greenish-white flash—came thrusting in.

After a little while I regained consciousness. Everything around me was pitch dark. Somehow I managed to figure out that I seemed to have been blown down the hall to the back part of the inn. I was buried under the wreckage of the two-story building and although I struggled to free myself and crawl out, I couldn't move my body. The great central pillar which always appeared in the proprietor's bragging about his inn was lying right in front of me. I had already decided I was going to die and closed my eyes and was half asleep when I heard my mother's voice calling my name over and over. Mother's calling voice brought me back to my senses and I opened my eyes and looked around and saw that the edge of the house which was pinning me down was beginning to burn and that the fire was coming toward me. I thought that if I didn't get out of there in a hurry I would be burned to death, and I called for Mother as loudly as I could. Mother pulled aside the boards and beams which were already on fire and pulled me out to safety. To this day I can't forget how happy I was at that moment. . . . But at once I was stunned by the completeness of the change which had taken place in my surroundings. Everything in sight which can be called a building is crushed to the ground and sending out flames. People who are burned so badly that the skin of their bodies is peeling off in red strips are raising shrieking cries that sound as though the victims would die the next minute. There are even some people who are already dead. The street is so covered with dead people and burned people stretched out and groaning, and with fallen houses and things, that we can't get through. While we are trying to think what to do next we notice that the flames are steadily coming closer to the west of us. I walked over the roof of a fallen house that wasn't burning yet and escaped in the direction where there were no flames.

I came out at the river bank on the shore. . . . There for the first time I realized that I had become separated from Mother. At the side of the Kyobashi River burned people were moaning, "Hot! Hot!" and jumping into the river, and since they could not move their bodies freely, they would call for help with the voices of those facing death, and then drown. The river became not a stream of flowing water but rather a stream of drifting dead bodies. No matter how much I might exaggerate the stories of the burned people who died shrieking and of how the city of Hiroshima was burned to the ground, the facts would still be clearly more terrible and I could never really express the truth on this piece of paper; on this point I ask for pardon. . . .

—*Girl, Junior College student.*—

Ah, that instant! I felt as though I had been struck on the back with something like a big hammer, and thrown into boiling oil. For some time I was unconscious. . . .

Through a darkness like the bottom of Hell I could hear the voices of the other students calling for their mother. I could barely sense the fact that the students seemed to be running away from that place. I immediately got up, and . . . just frantically ran in the direction they were all taking. . . . The place where I had been working was Tanaka-cho, a little more than 600 yards from the center of the explosion. . . .

At the base of the bridge, inside a big cistern that had been dug out there, was a mother weeping and holding above her head a naked baby that was burned bright red all over its body, and another mother was crying and sobbing as she gave her burned breast to her baby. In the cistern the students stood with only their heads above the water and their two hands, which they clasped as they imploringly cried and screamed, calling their parents. But every single person who passed was wounded, all of them, and there was no one to turn to for help. . . . I looked at my two hands and found them covered with blood. . . . Shocked, I put my hand into my *mompei*[3]

---

[3] A type of slacks worn by Japanese girls.

pocket to get out my handkerchief, but there was no handkerchief, nor pocket either. And my *mompei* were also burned off below my hips. I could feel my face gradually swelling up.... From inside the wreckage of the houses we would hear screaming voices calling "Help!" and then the flames would swallow up everything....

Even now the scars of those wounds remain over my whole body. On my head, my face, my arms, my legs and my chest. As I stroke those blackish-red raised scars on my arms, and every time I look in a mirror at this face of mine which is not like my face, and think that never again will I be able to see my former face and that I have to live my life forever in this condition, it becomes too sad to bear....

2. For a period of more than fifteen years after the end of World War II American and Japanese scientists worked together in an investigation of the medical effects of the atomic bomb on the people of Hiroshima. The U. S. Atomic Energy Commission published the results of this investigation in 1962. Here are some key excerpts:[4]

> The three main types of physical effects associated with a nuclear explosion, namely, blast and shock, thermal radiation, and nuclear radiation, each have the potentiality for causing death and injury to exposed persons....
>
> The frequency of burn injuries due to a nuclear explosion is exceptionally high. Most of these are flash burns caused by direct exposure to the thermal radiation, although individuals trapped by spreading fires may be subjected to flame burns. In addition, persons in buildings or tunnels close to ground zero[5] may be burned by hot gases and dust entering the structure even though they are shielded adequately from direct or scattered thermal radiation. Finally, mention must be made of the harmful effects of the nuclear radiations on the body. These represent a source of casualties entirely new to warfare....
>
> Some 95 percent of the population within a half mile from ground zero were casualties.... Beyond about 1.5 miles, however, the chances of survival were very greatly improved. Between 0.5 and 1.5 miles from ground zero a larger proportion of the population would probably have survived if immediate medical attention had been available....
>
> It was estimated that 20 to 30 percent of the fatal casualties in Hiroshima and Nagasaki were caused by flash burns. In the former city alone, some 40,000 fairly serious burn cases were reported. Apart from other injuries, thermal radiation burns would have been fatal to nearly all persons in the open, without appreciable protection, at distances up to 6,000 feet (1.1 miles) or more from ground zero. Even as far out as 12,000 to 14,000 feet (2.2 to 2.6 miles), there were instances of such burns which were bad enough to require treatment....
>
> There are a number of consequences of nuclear radiation which may not appear for some years after exposure. Among them, apart from genetic effects, are the formation of cataracts, nonspecific life shortening, leukemia,

---

[4]Samuel Glasstone, ed., *The Effects of Nuclear Explosions* (Washington, U.S. Atomic Energy Commission, 1962), pp. 547, 551, 565, 598, 599, 600, 601.
[5]"Ground zero" is the point on the ground directly below the explosion. At Hiroshima the bomb was dropped from an altitude of about 31,000 feet and timed to explode at about 2,000 feet, so that maximum effect was achieved and as little as possible of the bomb's energy was dissipated into the ground.

other forms of malignant disease, and retarded development of children *in utero*[6] at the time of exposure. . . .

The first definite evidence of an increase in the incidence of leukemia cases among the inhabitants of Hiroshima and Nagasaki was obtained in 1947. The peak apparently occurred between 1950 and 1952, and the incidence has subsequently decreased. . . .

A special research initiated in 1957 in Hiroshima, to compare the frequency of malignant neoplasms,[7] other than leukemia, in people exposed within about a mile of ground zero in 1945 with the incidence in unexposed populations, has yielded some interesting information. Although the study covers only a two-year period, the results suggest a two- to four-fold increase over the expected frequency for neoplastic disease in some organs (lung, stomach, breast, and ovary) in exposed persons. . . .

Among the mothers who were pregnant at the time of the nuclear explosions in Japan, and who received sufficiently large doses to show the usual radiation symptoms, there was marked increase over normal in the number of still-births and in the deaths of newly born and infant children. A study of the surviving children made 4 or 5 years later showed a slightly increased frequency of mental retardation. . . . Maldevelopment of the teeth, attributed to injury at the roots, was also noted in many of the children.

A comparison made about 1952 of exposed children, whose ages ranged from less than 1 to about 14 years at the time of the explosions, with unexposed children of the same age, showed that the former had somewhat lower average body weight and were less advanced in stature and sexual maturity. . . .

3. Physical damage was not the only kind suffered by the people of Hiroshima. This report was written in 1965, twenty years after the bombing, and appeared in *The New York Times Magazine:* [8]

The city government of Hiroshima issues a pocket medical-identification certificate to all people who lived through the atomic bombing or entered the radiation area within a couple of weeks after the bombing. The meaning of *"hibakusha,"* the word for these people, is somewhere between "survivor" and "sufferer." The booklet identifies its carriers as *hibakusha* (the "u" is not pronounced) and entitles them to free medical care. It is now held by 93,391 people. . . .

But there are many more survivors in Hiroshima who never have registered. They don't like to be identified; they talk of prejudice against them. . . .

[The] difference between the *hibakusha* and survivors of other disasters is that the sense of being apart—from the world, from the city—continues, with no terminal date in sight. . . . As the years go by and Hiroshima becomes ever more prosperous and thrusting, the gap never seems to close; if anything, it seems to widen a bit. . .

Although not all *hibakusha* are poor, as a group they are economically lower than the rest of the city. Financial and economic power in Hiroshima

---

[6]Children about to be born.
[7]Cancers and tumors.
[8]A. M. Rosenthal, "The Taste of Life in Hiroshima Now," *The New York Times Magazine* (August 1, 1965), pp. 30, 32-35.

is held largely by newcomers, and the proportion of *hibakusha* on the relief rolls is perhaps twice as high as it is in the city's population....

"Generally speaking, the *hibakusha* are physically weak and cannot undertake the same labor as the new people." This was said by Mayor Hamai; it was also said by doctors, businessmen, employers, shopkeepers. Everybody believes it....

Over and over you hear from *hibakusha*—"I cannot work hard," "I get tired quickly; what can I do?" "We know we cannot produce as much." Over and over, everywhere. Hiroshima, the new Hiroshima, talks of work and new buildings and "hot money"; the *hibakusha* talk of fatigue and listlessness....

"*Hibakusha* live with the belief that they are weaker, unlucky and not able, and this itself is a real handicap to them in their lives," said a Japanese physician....

Almost every time *hibakusha* discuss their problems or attitudes toward them, the question of marriage and the prejudice against marrying somebody who may have "bad seed" as the result of radiation comes up. A doctor at the Atomic Bomb Hospital tells about a girl survivor who went to Tokyo, married, had two children and only then told her husband the truth—that she had not been outside Hiroshima that Aug. 6. "With two children, he wouldn't divorce me," she told the doctor.

But everything is not all right for other young *hibakusha*. Mayor Hamai tells of men and women who refuse to apply for survivor medical benefits because they are of marriageable age and don't want to be registered, and known, as *hibakusha*.

And the survivors seem to be afflicted by something deeper than hypochondria, deeper than lethargy. It seems to be a sense of being hostage, of having had doom ordained that day and of waiting for it to catch up....

One of the few psychological studies of the *hibakusha* was carried out by Dr. Lifton,[9] based on long interviews over a period of four and a half months. He found some "distinct psychological elements": a feeling of "continuous encounter with death," a "loss of faith (or trust) in the structure of existence," an "overwhelming encounter with death," "psychologically speaking, no end point, no resolution."

"This continuous and unresolvable encounter with death, then, is a unique feature of the atomic-bomb disaster," he wrote.

Dr. Lifton called his report "The Theme of Death," and it is a theme still heard against the noise of the trucks, the piledriver and the cashbox....

B. THE REASONS WHY

Aside from the President, the man who exercised the greatest influence in the shaping of the decision to use the bomb was Secretary of War Henry L. Stimson. Stimson had a long record of distinguished public service. A lifelong Republican and Secretary of State under Herbert Hoover from 1929 to 1933, he had been appointed Secretary of War in 1940 by Democratic President Franklin D. Roosevelt in an effort to ensure bipartisan support for the impending war effort. Stimson served throughout the war, and was one of the few political figures who participated in every phase of the planning and direction of the atomic bomb project.

---

[9]Dr. Robert J. Lifton, American psychiatrist.

1. **Stimson wrote an authoritative analysis of the reasoning behind the American decision in an article published two years after Hiroshima, from which the following has been excerpted:**[10]

> The principal political, social, and military objective of the United States in the summer of 1945 was the prompt and complete surrender of Japan. Only the complete destruction of her military power could open the way to lasting peace.
>
> Japan, in July, 1945, had been seriously weakened by our increasingly violent attacks.... There was as yet no indication of any weakening in the Japanese determination to fight rather than accept unconditional surrender. If she should persist in her fight to the end, she had still a great military force.
>
> In the middle of July 1945, the intelligence section of the War Department General Staff estimated Japanese military strength as follows: in the home islands, slightly under 2,000,000; in Korea, Manchuria, China proper, and Formosa, slightly over 2,000,000; in French Indo-China, Thailand, and Burma, over 200,000; in the East Indies area, including the Philippines, over 500,000; in the by-passed Pacific islands, over 100,000. The total strength of the Japanese army was estimated at about 5,000,000 men. These estimates later proved to be in very close agreement with official Japanese figures.
>
> The Japanese Army was in much better condition than the Japanese Navy and Air Force. The Navy had practically ceased to exist except as a harrying force against an invasion fleet. The Air Force had been reduced mainly to reliance upon Kamikaze, or suicide, attacks. These latter, however, had already inflicted serious damage on our seagoing forces, and their possible effectiveness in a last ditch fight was a matter of real concern to our naval leaders.
>
> As we understood it in July, there was a very strong possibility that the Japanese Government might determine upon resistance to the end, in all the areas of the Far East under its control. In such an event the Allies would be faced with the enormous task of destroying an armed force of five million men and five thousand suicide aircraft, belonging to a race which had already amply demonstrated its ability to fight literally to the death.
>
> The strategic plans of our armed forces for the defeat of Japan, as they stood in July, had been prepared without reliance upon the atomic bomb, which had not yet been tested in New Mexico. We were planning an intensified sea and air blockade, and greatly intensified strategic air bombing, through the summer and early fall, to be followed on November 1 by an invasion of the southern island of Kyushu. This would be followed in turn by an invasion of the main island of Honshu in the spring of 1946. The total U. S. military and naval force involved in this grand design was of the order of 5,000,000 men; if all those indirectly concerned are included, it was larger still.
>
> We estimated that if we should be forced to carry this plan to its conclusion, the major fighting would not end until the latter part of 1946, at the earliest. I was informed that such operations might be expected to cost over

---

[10]Henry L. Stimson, "The Decision to Use the Atomic Bomb," *Harpers* (February, 1947), pp. 101-102, 105-107.

a million casualties, to American forces alone. Additional large losses might be expected among our allies and, of course, if our campaign were successful and if we could judge by previous experience, enemy casualties would be much larger than our own....

The New Mexico test occurred while we were at Potsdam, on July 16. It was immediately clear that the power of the bomb measured up to our highest estimates. We had developed a weapon of such a revolutionary character that its use against the enemy might well be expected to produce exactly the kind of shock on the Japanese ruling oligarchy which we desired, strengthening the position of those who wished peace, and weakening that of the military party....

Hiroshima was bombed on August 6, and Nagasaki on August 9. These two cities were active working parts of the Japanese war effort. One was an army center; the other was naval and industrial. Hiroshima was the headquarters of the Japanese Army defending southern Japan and was a major military storage and assembly point.... We believed that our attacks had struck cities which must certainly be important to the Japanese military leaders, both Army and Navy, and we waited for a result. We waited one day....

After a prolonged Japanese Cabinet session in which the deadlock was broken by the Emperor himself, the offer to surrender was made on August 10.... Our great objective was thus achieved, and all the evidence I have seen indicates that the controlling factor in the final Japanese decision to accept our terms of surrender was the atomic bomb....

Two great nations were approaching contact in a fight to a finish which would begin on November 1, 1945. Our enemy, Japan, commanded forces of somewhat over 5,000,000 armed men. Men of these armies had already inflicted upon us, in our breakthrough of the outer perimeter of their defenses, over 300,000 battle casualties. Enemy armies still unbeaten had the strength to cost us a million more. *As long as the Japanese Government refused to surrender,* we should be forced to take and hold the ground, and smash the Japanese ground armies, by close-in fighting of the same desperate and costly kind that we had faced in the Pacific Islands for nearly four years....

In order to end the war in the shortest possible time and to avoid the enormous losses of human life which otherwise confronted us, I felt that we must use the Emperor as our instrument to command and compel his people to cease fighting ... and that to accomplish this we must give him and his controlling advisers a compelling reason to accede to our demands. This reason furthermore must be of such a nature that his people could understand his decision. The bomb seemed to me to furnish a unique instrument for that purpose.

My chief purpose was to end the war in victory with the least possible cost in the lives of the men in the armies which I had helped to raise. In the light of the alternatives which, on a fair estimate, were open to us I believe that no man, in our position and subject to our responsibilities, holding in his hand a weapon of such possibilities for accomplishing this purpose and saving those lives, could have failed to use it and afterwards looked his countrymen in the face....

The face of war is the face of death; death is an inevitable part of every order that a wartime leader gives. The decision to use the atomic bomb was a decision that brought death to over a hundred thousand Japanese. No explanation can change that fact and I do not wish to gloss it over. But this deliberate, premeditated destruction was our least abhorrent choice. The destruction of Hiroshima and Nagasaki put an end to the Japanese war. It stopped the fire raids, and the strangling blockade; it ended the ghastly specter of a clash of great land armies. . . .

SECTION 2

Was It a Military Decision?

World War II was history's first truly global conflict. Virtually all the continents were involved, and its battles raged across all the oceans. Yet it was one indivisible struggle. Decisions concerning one theater of operations affected and were affected by decisions in every other theater. This was as true of the decision to use the bomb as of any other.

A. SETTING THE STAGE FOR THE GREAT DECISION

The materials that follow present certain military and political complexities of World War II in all their interrelatedness. They comprise the historical context within which the dilemmas of 1945 had to be resolved.

1. A selection of the war's principal events appears in the following chronology. Many of these developments may seem remote from the atomic bomb decision. The fact is, however, that each of them influenced it in some way.

World War II: A Selected Chronology

	Military Events	Political Events
1939		
Sept.	Germany invades Poland.	Britain, France declare war on Germany. World War II begins. U.S. proclaims neutrality.
1940		
April–June	Germany conquers Norway, Denmark, Holland, Belgium, France, and launches air assault on Britain.	
Sept.	Japan, having already conquered much of China, moves into French Indo-China.	First peacetime draft law enacted in U.S.
Nov.		Roosevelt reelected President for third term.

12

1941

March		Lend-Lease Act ensures U.S. aid to Britain.
June	Germany invades Soviet Union.	U.S. extends aid to Soviet Union.
July-Dec.		U.S. pressures Japan to end aggression in China, Indo-China.
Dec.	Japanese surprise attack on Pearl Harbor cripples U.S. Pacific fleet.	U.S. declares war on Japan. Germany and Italy declare war on U.S.

1942

Jan.-May	Japan conquers Philippines, Guam, Wake Island, Malaya, Singapore, Hong Kong, Dutch East Indies, and vast areas of western Pacific.	
May-June	U.S. defeats Japan in naval battles of Coral Sea and Midway.	
Aug.	U.S. starts first land offensive against Japan on Guadalcanal Island.	
Oct.-Dec.	Soviet counteroffensive at Stalingrad, U.S. and British offensives in North Africa, turn tide against Germany.	

1943

June	U.S. launches "island-hopping" campaign towards Japan.	

1944

June	"D-Day": huge Anglo-American invasion of France opens 2nd front against Germany.	
Nov.		Roosevelt reelected for fourth term. Harry S. Truman elected Vice President.
Dec.	By year's end U.S. Pacific campaign reaches Philippines and Marianas, within bombing range of Japan.	

1945

Feb.		Yalta Conference. Roosevelt, Churchill, Stalin agree on postwar Europe. U.S.S.R. agrees to enter war against Japan within 90 days after Germany surrenders.
Feb.-March	Bloody Battle of Iwo Jima brings U.S. to within 750 miles of Tokyo.	
April		Roosevelt dies. Truman accedes to the Presidency.
April-June	U.S. conquest of Okinawa shortens distance to 360 miles, but costs heavily in casualties.	
May		Germany surrenders.
July		Potsdam Conference. Truman, Churchill, Stalin issue Potsdam Declaration demanding Japan's "unconditional surrender." U.S.S.R. confirms agreement to enter war in August.
Aug. 6	U.S. drops first atomic bomb on Hiroshima.	
Aug. 8	U.S.S.R. enters war against Japan.	
Aug. 9	U.S. drops second atomic bomb on Nagasaki.	
Aug. 14		Japan accepts terms of Potsdam Declaration. World War II ends.

2. Almost at once after America's entry into the war in December 1941, a crucial strategic choice had to be made. Should the main American effort be directed against Germany or Japan? After heated debate, it was decided that Germany was the more dangerous enemy and must be crushed first. In the meantime, only limited operations could be carried out against Japan.

This strategy produced the "island-hopping" campaign in the Pacific theater. Step by step for three years the Americans fought their way across the Pacific in a two-pronged offensive converging on Japan from Hawaii and Australia, using the Japanese-occupied islands of the south and central Pacific as stepping-stones. Compared with the huge armies engaged in the European theater, the battles for these islands involved relatively small numbers of men, but they produced some of the most vicious fighting in all the annals of warfare. This agonizing experience, and the grim prospect of what might still lie ahead, bore heavily upon American leaders in 1945 as they sought the best way to end the war.

Early in 1945 a tiny volcanic island named Iwo Jima, only 750 miles south of Tokyo, was selected as the target for the next step of the American island-hopping campaign. The result was the costliest battle in 168 years of U. S. Marine Corps history. Nearly

20,000 men of the two attacking Marine divisions became casualties in less than one month of fighting.

The following description was written soon after the battle by a war correspondent who landed with the Marines at Iwo Jima.[1]

> Whether the dead were Japs or Americans, they had one thing in common; they had died with the greatest possible violence. Nowhere in the Pacific war had I seen such badly mangled bodies. Many were cut squarely in half. Legs and arms lay fifty feet away from any body. In one spot on the sand, far from the nearest cluster of dead, I saw a string of guts 15 feet long. Only legs were easy to identify; they were Jap if wrapped in khaki puttees, American if covered by canvas leggings. The smell of burning flesh was heavy in some areas.
>
> What the Japs succeeded in doing was this: They built underground so well that they all but nullified our superior firepower. We could bomb and shell until our guns sizzled and our pilots dropped.... But when our barrage lifted and our infantry advanced the Japs were back in position, firing their machine guns and mortars.
>
> At times it was agonizing to realize that we progressed so slowly and at so high a price, in spite of our superior strength. For all our technical skill, we had on Iwo no method and no weapon to counteract the enemy's underground defense. The Japs made us fight on their own terms. We could beat them on their own terms; we could kill four for one; we could make them take as many casualties as we took (and their casualties were nearly all dead). But the men fighting on Iwo Jima frankly thought that price too high. The Japs didn't seem to mind dying; we preferred to live....
>
> The hopes we had of a quick victory melted away slowly. One day it seemed that only a few more days would be required; next day it seemed that surely a break would come somewhere; a week later our progress was still being measured 50 yards, 100 yards, 300 yards, at a time.... [The Japs] stayed in their tunnels and their molehills to the deathly end, and we had to go in and dig them out or burn [them] out or seal them in. There was nothing else for us to do. It was brave men against Jap cunning and Jap steel.

3. The following account of the battle for Iwo Jima and its repercussions on the home front has been excerpted from a book published twenty years later by a U.S. Navy war correspondent:[2]

> The [American] tanks, now maneuvering inland, were helping, but there was a surprising amount of opposition left in the neck of the island. For days it had been saturated with naval and aerial bombs. Still the enemy lived on. Captain Masao Hayauchi's 12th Independent Anti-Tank Battalion knocked out several Shermans [American tanks]. When his guns were smashed, he led the last assault. Clutching to his chest a charge with the fuse lit, he splayed himself against a tank and blew himself up....
>
> Private First Class George Smyth, eighteen, of Brooklyn, had never seen such Japanese. They were 6-footers, and they never retreated. Smyth's

---
[1] Robert Lee Sherrod, *On To Westward: War in the Central Pacific* (New York, Duell, Sloan & Pearce, 1945), pp. 180, 202-203.
[2] Richard F. Newcomb, *Iwo Jima* (Holt, Rinehart and Winston, and McIntosh and Otis Inc., copyright (c) 1965 by Richard F. Newcomb), pp. 118, 174, 176, 236-237, 241, 252.

buddy fell beside him, a pistol bullet through his head, dead center. It came from a captured Marine .45. On the other side, a Japanese came down with his sword, both hands grasping the hilt. The Marine put up his right hand to ward off the blow, and his arm was sliced down the middle, fingers to elbow. As Smyth ran forward, a Japanese disappeared before him into a hole. Smyth dropped at the hole to finish him off, but the Japanese was already rising from a tunnel behind him. Smyth turned just in time to kill him. The ground was giving . . . [the enemy] every advantage, and they were using them all. By nightfall, nearly every [American] gain in the center had been nullified. . . .

The 26th [Marine regiment] made nearly 200 yards, capturing a strongpoint that had held up the advance the previous day, and found itself in front of a knoll, north of what had been Nishi village. The enemy fell strangely silent, and the Marines cautiously surrounded the hill. Demolitions men blasted one cave entrance closed, and machine gunners cut down Japanese who ran from a rear entrance. The hill still looked suspicious, but Marines ran to the top of it. Just then the whole hill shuddered and the top blew out with a roar heard all over the island. Men were thrown into the air, and those nearby were stunned by concussion. Dozens of Marines disappeared in the blast crater, and their comrades ran in to dig for them. Strong men vomited at the sight of charred bodies, and others walked from the area crying. The enemy had blown up his own command post, inflicting forty-three Marine casualties at the same time. . . .

On February 27 the subject [of American losses] sprang into national prominence. The *San Francisco Examiner,* in a frontpage editorial, said that while the Marines would no doubt capture Iwo Jima, "there is awesome evidence in the situation that the attacking American forces are paying heavily for the island, perhaps too heavily."

"It is the same thing that happened at Tarawa and Saipan . . ." the editorial said, "and if it continues the American forces are in danger of being worn out before they ever reach the really critical Japanese areas." . . .

On March 16, the Navy disclosed that it had received "a number" of letters, and released one as typical.

A woman wrote:

"Please, for God's sake, stop sending our finest youth to be murdered on places like Iwo Jima. It is too much for boys to stand, too much for mothers and homes to take. It is driving some mothers crazy. Why can't objectives be accomplished in some other way? It is almost inhuman and awful—stop, stop!"

The Navy said that Secretary Forrestal had replied: . . .

"[We] have now, no final means of winning battles except through the valor of the Marine or Army soldier who, with rifle and grenade, storms enemy positions, takes them, and holds them. There is no shortcut or easy way. I wish there were."

4. Iwo Jima was followed by the battle of Okinawa (April-June 1945), which was planned as the last step before the expected invasion of Japan. The following excerpt taken from the official U.S. Army history suggests the nature of the fighting there:[3]

---

[3]Roy E. Appleman, James M. Burns, Russell A. Gugeler, John Stevens, *Okinawa: The Last Battle, United States Army in World War II* (Washington, Government Printing Office, 1948) pp. 384-386.

Nothing illustrates so well the great difference between the fighting in the Pacific and that in Europe as the small number of military prisoners taken on Okinawa. At the end of May the III Amphibious Corps had captured only 128 Japanese soldiers. At the same time, after two months of fighting in southern Okinawa, the four divisions of the XXIV Corps had taken only 90 military prisoners. The 77th Division, which had been in the center of the line . . . had taken only 9 during all that time. Most of the enemy taken prisoner either were badly wounded or were unconscious; they could not prevent capture or commit suicide before falling into American hands.

In the light of these prisoner figures there is no question as to the state of Japanese morale. The Japanese soldier fought until he was killed. There was only one kind of Japanese casualty—the dead. . . .

Casualties on the American side were the heaviest of the Pacific war. . . .

Nonbattle casualties were numerous, a large percentage of them being neuropsychiatric or "combat fatigue" cases. . . . The most important cause of this was unquestionably the great amount of enemy artillery and mortar fire, the heaviest concentrations experienced in the Pacific War. Another cause of men's nerves giving way was the unending close-in battle with a fanatical foe. The rate of psychiatric cases was probably higher in Okinawa than in any previous operation in the Pacific.

5. A U.S. Naval Academy historian and the commander of American naval forces in the Pacific describe the effect of Japanese suicide planes (Kamikaze) on the U.S. fleet at Okinawa:[4]

From the beginning, Japanese bombers and suicide planes made sporadic attacks on the American ships off Okinawa. On March 31 a kamikaze crashed into Spruance's flagship *Indianapolis,* releasing a bomb that penetrated several decks and blew two holes in her hull. While Spruance transferred his flag to the old battleship *New Mexico,* the *Indianapolis* was patched up in the Kerama anchorage and then headed for Mare Island Navy Yard for extensive repairs. On April 4 a crashing kamikaze so mangled a destroyer-transport that she had to be sunk. By April 5 Japanese bombers and suicide planes had succeeded in damaging 39 naval vessels, including two old battleships, three cruisers, and an escort carrier. These raids however were mere preliminaries to the general counterattack which the Imperial Army and Navy, acting for the first time in really close concert, launched on April 6. On the morning of the 6th a Japanese reconnaissance plane sighted TF 58 east of Okinawa. Shortly afterward 355 kamikaze pilots in old aircraft rigged for suicide attack began taking off from airfields in Kyushu. Some headed for TF 58, others for the shipping off Okinawa.

First and most persistently attacked by the kamikazes were the outlying picket vessels, which early in the campaign generally had only their own guns to protect themselves. In mid-afternoon of the 6th, suicide planes swarmed down on the destroyer *Bush* on picket patrol north of Okinawa and made three hits. The destroyer *Colhoun,* patrolling the adjacent station, rushed to support the damaged *Bush* and was herself crashed by three kamikazes. Both destroyers began to sink. An alert combat air patrol and long-practiced countermeasures prevented the enemy aircraft from reach-

---

[4]E. B. Potter and Chester W. Nimitz, eds., *The Great Sea War: The Story of Naval Action in World War II,* (c) 1962 by Prentice-Hall, Inc., pp. 452, 453, 454, 455.

ing TF 58 that day, but about 200 reached the Okinawa area. Here most of the attackers were disposed of by fighter planes and by antiaircraft fire so intense that a hail of falling shell fragments caused 38 American casualties. Nevertheless the enemy planes damaged 22 naval vessels, sank a destroyer-transport and an LST, and demolished two loaded ammunition ships, leaving the Tenth Army short of certain types of shell. . . .

On April 7 a Kamikaze at last penetrated the TF 58 air patrol and crashed into the deck of the carrier *Hancock,* killing 43 men. By nightfall, suicide planes had damaged four more naval vessels. The April 6-7 raid was only the first of ten general kamikaze attacks launched against the fleet and shipping off Okinawa. Smaller-scale suicide and conventional air raids occurred nearly every day. . . .

Though the kamikazes continued to find most of their victims among the radar pickets and the ships off Okinawa, the fast carrier force took its share of hits. Admiral Mitscher lost a large part of his staff and had to shift flagships twice in three days as the carriers *Bunker Hill* and *Enterprise* were successively hit and put out of action by crashing kamikazes. Southeast of Okinawa the British task force also came under persistent kamikaze attack. All four British carriers were hit, but all were able to continue operations. . . .

Nearly 13,000 Americans had been killed, of whom 3,400 were marines and 4,900 navy. In the fleet most of the casualties among ships and men were the result of enemy air attack, chiefly by suicide planes. By air attack alone 15 naval vessels were sunk, none larger than a destroyer, and more than 200 were damaged, some beyond salvage. This costly sacrifice had purchased a position for bringing air power to bear heavily upon the industrial centers of southern Japan and a base for completing the blockade of the home islands and for supporting an invasion of Kyushu.

B. THE NEW COMMANDER-IN-CHIEF

Franklin D. Roosevelt died suddenly on April 12, 1945, after twelve years as President of the United States. That evening Harry S. Truman was sworn in as the new President. Truman had been a respected but little known Senator from Missouri until the election campaign of 1944, barely six months before. Then Roosevelt unexpectedly discarded Henry A. Wallace, his Vice President during the preceding four years, and selected Truman to run in his place.

Through no fault of his own, Truman was singularly ill-prepared to take over the reins that Roosevelt had held so confidently during the long and tumultuous years of his administration. Forty-eight hours after the inauguration of the new team on January 20, Roosevelt had left for the Yalta Conference. He was out of the country for more than six weeks, and when he returned he was visibly weary and ill. As a result Truman was never adequately briefed on many vital matters of state.

1. Truman had never even been told that an atomic bomb was being developed. How he learned of it is described in the following excerpt:[5]

> Harry Truman got his first inkling of the decision that would be his only an hour after becoming President of the United States on the night of April 12. He took the oath at 7:09 p.m. and held a brief cabinet meeting to assume

---
[5] Fletcher Knebel and Charles W. Bailey II, *No High Ground* (New York, Harper & Row, 1960), pp. 99-100.

the responsibility of office. As the Roosevelt cabinet members filed silently out, Stimson remained behind. He told the new President that the government he now headed was developing a weapon of enormous power. He provided no details. The next day, James F. Byrnes[6] . . . told the President more of the awful force of the proposed bomb, but not until Vannevar Bush[7] called a few days later did Truman get an extensive scientific explanation.

2. What kind of man was the new President? A portrayal of Harry Truman as he seemed to many upon taking office was drawn by historian Eric F. Goldman in his book about the postwar era, *The Crucial Decade.* [8]

>   Life had prepared Harry Truman for the simple. Life was Jackson County, Missouri, where you grew up near-sighted and the other boys snickered, you went courting Bess Wallace in the big house and her parents tried to get rid of you, you came back from the war to start the haberdashery with Eddie Jacobson and the store went broke. But you stuck at it. . . .
>
>   Despite the uppity parents, you married Bess and got started in politics under the wing of Tom Pendergast. The "damned New York liberals" might keep attacking Pendergast as a corrupt Boss. Did your friend Tom ever ask you to do anything dishonest? And it was quite a life, this being a County Judge, then United States Senator, riding off to the Capital and coming back to Independence with people gathering around and asking, "Harry, what are they up to in Washington?"
>
>   "They're up to getting whatever they can get," you would say. Everybody would laugh and there was a night of bourbon and poker with the boys and a long visit with Mother, who would always send you off with "Now, Harry, hew to the good line." You hewed to the good line. You never went near a tainted dime. You were a loyal Democrat, almost always voting for the New Deal laws proposed by the Democratic President. You liked reading Civil War history and you remembered World War I; this time no smart-aleck businessmen were going to make off with huge war profits. So you stood up from your back bench in the Senate and called for a committee to keep a suspicious eye on the whole hurry and grab. Naturally you were chairman and naturally you worked nights watching over every dollar spent and soon all kinds of people had a good word for Harry Truman's Special Committee Investigating the National Defense Program.
>
>   You kept holding to the simple rules and good things went right on happening. 1944 and party leaders were telling Franklin Roosevelt that Henry Wallace simply would not do. Why not someone who had voted New Deal yet was not too much stamped as a New Dealer, a friendly man who had stirred up few enmities but nevertheless had some national standing, a Senator from the Border States which could easily decide the election? . . .
>
>   The afternoon it happened [the death of Roosevelt] you were having a drink in Sam Rayburn's office. You turned ashen and your voice stuck in your throat. The next day you said to the reporters: "Boys, if you ever pray, pray for me now. I don't know whether you fellows ever had a load of hay

---

[6]Truman had summoned Byrnes on the first morning of his Presidency to offer him the post of Secretary of State. Byrnes had been Director of the Office of War Mobilization throughout much of the course of the war, and in that position had been in on the development of the bomb project.

[7]Bush was Director of the Office of Scientific Research and Development, and as such the top-ranking scientist in the government.

[8]Eric F. Goldman, *The Crucial Decade: America, 1945-1955* (New York, Alfred A. Knopf, 1956), pp. 16-19.

fall on you, but when they told me yesterday what had happened, I felt like the moon, the stars, and all the planets had fallen on me."

The first weeks in the White House were awful. You had never really wanted this. Sure, you were a good man, good as the next one, and there was no reason why you shouldn't be Vice-President. But that didn't mean you should be President of the United States, especially not after Franklin Roosevelt and in times like these. His shadow was always over your shoulder. You had been let in so little on really important affairs that you had to keep summoning Roosevelt intimates . . . merely to get the basic facts. And every day there was some tremendous decision to make in that lonely room where Abraham Lincoln and Woodrow Wilson and Franklin Roosevelt had sat.

C. ADVICE FROM THE MILITARY

In the spring and summer of 1945, as the war approached its final stages, a momentous controversy boiled up among America's military leaders. The Army on one side, and the Navy and Air Force on the other, put forth opposing views as to the best way to defeat Japan. The question as to whether or not we would use the atomic bomb depended on the outcome of this dispute.

1. On June 18, 1945, President Truman held a meeting with his top military aides at which crucial decisions were made. The Army's point of view was expounded by General George C. Marshall, and later reported in official government records. Marshall was a distinguished soldier who had served for over forty years. Appointed Chief of Staff in 1939, he had served in that key post throughout the war. He and his civilian superior, Secretary of War Stimson, tended to agree on most matters—as they did on this one:[9]

> General Marshall said that it was his personal view that the operation against Kyushu was the only course to pursue. He felt that air power alone was not sufficient to put the Japanese out of the war. It was unable to put the Germans out . . . Against the Japanese, scattered through mountainous country, the problem would be much more difficult than it had been against Germany. He felt that this plan offered the only way the Japanese could be forced into a feeling of utter helplessness.

2. The views of Stimson and Marshall were reinforced by the recommendations of the Army's brilliant field commander in the Pacific, General Douglas MacArthur. In his memoirs, published in 1964, MacArthur described the strategy he urged upon his superiors:[10]

> On April 12th, General Marshall asked my views as to future Pacific operations. . . .
> I replied on April 20th strongly recommending a direct attack on the Japanese mainland at Kyushu for the purpose of securing airfields to cover the main assault on Honshu. . . . I recommended a target date of November 1st.

---

[9] U.S. Department of State, *Foreign Relations of the United States, Diplomatic Papers: The Conference of Berlin (Potsdam) 1945* (Washington, Government Printing Office, 1960), I, 906. [Hereafter referred to as *Potsdam Papers.*]

[10] Douglas MacArthur, *Reminiscences* (New York, McGraw-Hill, 1964), pp. 260-261.

3. Senior officers of the U.S. Navy dissented from the recommendations advanced by Stimson, Marshall, and MacArthur. Admiral William D. Leahy had been Chief of Naval Operations before the war, and served from 1942 to 1945 as personal military advisor to Presidents Roosevelt and Truman. In his memoirs, published in 1950, Leahy states:[11]

> By the beginning of September [1944], Japan was almost defeated through a practically complete sea and air blockade. However, a proposal was made by the Army to force a surrender of Japan by an amphibious invasion of the main islands through the Island of Kyushu. This was discussed at length by the Joint Chiefs of Staff but final decision was not reached.
>
> The JCS did authorize the preparation of plans for an invasion, but the invasion itself was never approved. The Army did not appear to be able to understand that the Navy, with some Army air assistance, already had defeated Japan. The Army not only was planning a huge land invasion of Japan, but was convinced that we needed Russian assistance as well to bring the war against Japan to a successful conclusion.
>
> It did not appear to me that under the then existing conditions there was any necessity for the great expenditure of life involved in a ground force attack on the numerically superior Japanese Army in its home territory. My conclusion, with which the naval representatives agreed, was that America's least expensive course of action was to continue and intensify the air and sea blockade and at the same time to occupy the Philippines.
>
> I believed that a completely blockaded Japan would then fall by its own weight. Consensus of opinion of the Chiefs of Staff supported this proposed strategy, and President Roosevelt approved....
>
> [Under the pressure of events during the next nine months, however, this consensus broke up.]
>
> A conference was held at the White House [on June 18, 1945] primarily to discuss the necessity and practicability of invading the Japanese home islands....
>
> At the White House session General Marshall and Admiral King both strongly advocated an invasion of Kyushu at the earliest possible date. This was a modification of King's stand. Until then the Admiral had preferred an invasion of the coast of China, possibly in the Amoy area. King thought that would be a good place to prepare for a major operation on the Japanese mainland. He had never been as positively opposed to invasion as I had. Either operation—Kyushu or Amoy—would be difficult and hazardous, and apparently he had decided to go along with Marshall in proposing Kyushu.
>
> General Marshall was of the opinion that such an effort would not cost us in casualties more than 63,000 of the 190,000 combatant troops estimated as necessary for the operation.
>
> The President [Truman] approved the Kyushu operation and withheld for later consideration a general invasion of Japan. The Army seemed determined to occupy and govern Japan by military government as was being done in Germany. I was unable to see any justification, from a national defense point of view, for an invasion of an already thoroughly

---

[11] William D. Leahy, *I Was There* (New York, Whittlesey House, 1950, pp. 259, 384, 385, 441. By permission of Brandt & Brandt.

defeated Japan. I feared that the cost would be enormous in both lives and treasure.

It was my opinion at that time that a surrender could be arranged with terms acceptable to Japan that would make fully satisfactory provisions for America's defense against any future trans-Pacific aggression.

To some extent I was going counter to the principle of unconditional surrender. However, at Casablanca and subsequent meetings we had not agreed with anybody to demand an unconditional surrender of Japan. That policy had been approved only as it applied to Europe.

Naturally, I had acquainted President Truman with my own ideas about the best course to pursue in defeating Japan as fully as I had done with President Roosevelt. Truman was always a good listener, and I could not gauge exactly what his own feeling was. He did indicate in our discussion that he was completely favorable toward defeating our Far Eastern enemy with the smallest possible loss of American lives. It wasn't a matter of dollars. It might require more time—and more dollars—if we did not invade Japan. But it would cost *fewer lives*....

In the spring of 1945 President Truman directed Mr. Byrnes to make a special study of the status and prospects of the new atomic explosive on which two billion dollars already had been spent. Byrnes came to my home on the evening of June 4 to discuss his findings. He was more favorably impressed than I had been up to that time with the prospects of success in the final development and use of the new weapon.

Once it had been tested, President Truman faced the decision as to whether to use it. He did not like the idea, but was persuaded that it would shorten the war against Japan and save American lives. It is my opinion that the use of this barbarous weapon at Hiroshima and Nagasaki was of no material assistance in our war against Japan. The Japanese were already defeated and ready to surrender because of the effective sea blockade and the successful bombing with conventional weapons.

4. If, as Leahy claims, the victory over Japan was primarily a naval victory, then no man bore a greater responsibility for it than Admiral Ernest J. King, Commander-in-Chief of the U.S. Fleet and Chief of Naval Operations from 1941 to 1945. In his memoirs King sheds a revealing light upon the decision-making process:[12]

Upon Marshall's insistence, which also reflected MacArthur's views, the Joint Chiefs had prepared plans for landings in Kyushu and eventually in the Tokyo plain. King and Leahy did not like the idea; but as unanimous decisions were necessary in the Joint Chiefs meetings, they reluctantly acquiesced, feeling that in the end sea power would accomplish the defeat of Japan, as proved to be the case....

The first successful atomic explosion took place at Alamagordo, New Mexico, on 16 July 1945.... President Truman gave Secretary Stimson the go-ahead signal for the use of the atomic bomb.... The President in giving his approval for these attacks appeared to believe that many thousands of American troops would be killed in invading Japan, and in this he was entirely correct; but King felt, as he had pointed out many times, that the dilemma was an unnecessary one, for had we been willing to wait, the

---

[12]Ernest J. King and Walter M. Whitehill, *Fleet Admiral King: A Naval Record* (New York, W. W. Norton, Copyright 1952), pp. 598, 621.

effective naval blockade would, in the course of time, have starved the Japanese into submission through lack of oil, rice, medicines, and other essential materials. The Army, however, with its underestimation of sea power, had insisted upon a direct invasion and an occupational conquest of Japan proper. King still believes this was wrong.

5. General H. H. Arnold commanded the Army Air Force in World War II. Not surprisingly, his reasons for disagreeing with the Army's views, as stated in his postwar memoirs, are different from those of the naval commanders:[13]

> As soon as I could, [in July, 1945] I submitted a paper to the Joint Chiefs of Staff asking that the concept for employment of Air against Japan be changed....
>
>> I consider that our concept for operation against Japan should be to place, initially, complete emphasis upon strategic Air offensive, complemented by a Naval and Air blockade.... Estimates of the Joint Target Group indicate that the military and economic capacity of the Japanese nation can be destroyed by an effective dropping, on Japan, of 1,600,000 tons of bombs. This tonnage should disrupt industry, paralyze transportation and seriously strain the production and distribution of foods and other essentials of life. These effects might cause a capitulation of the enemy....
>
> The surrender of Japan was not entirely the result of the two atomic bombs. We had hit some 60 Japanese cities with our regular H. E. [High Explosive] and incendiary bombs, and as a result of our raids, about 241,000 people had been killed, 313,000 wounded, and about 2,333,000 homes destroyed. Our B-29's had destroyed most of the Japanese industries and, with the laying of mines, which prevented the arrival of incoming cargoes of critical items, had made it impossible to carry on a large-scale war.... Accordingly, it always appeared to us that, atomic bomb or no atomic bomb, the Japanese were already on the verge of collapse.

D. WAS THERE A DECISION AT ALL?

There is a central mystery about the decision to use the atomic bomb. The testimony of the two documents that follow, one by President Truman and the other by British Prime Minister Winston Churchill, indicates its nature.

1. According to the United States Constitution, the ultimate power of military decision rests with the President in his capacity as Commander-in-Chief of the armed forces. Although he relied heavily on the advice of Stimson and Marshall, Truman assumed full responsibility for the use of the bomb. In a letter written in 1953 in his typically emphatic manner, Truman reviewed what happened when news of the first successful atomic explosion reached him at the Potsdam Conference, near Berlin, where he was engaged in a summit meeting with Prime Minister Churchill and Premier Stalin of the Soviet Union:[14]

---

[13]H. H. Arnold, *Global Mission* (New York, Harper and Row, copyright 1949), pp. 595, 596, 598.

[14]Letter, Truman to Prof. J. L. Cate, January 12, 1953, cited in Wesley F. Craven and James L. Cate, eds., *The Army Air Forces in World War II* (Chicago, University of Chicago Press, 1953), V, insert between pp. 712 and 713.

When the message came to Potsdam that a successful atomic explosion had taken place in New Mexico, there was much excitement and conversation about the effect on the war then in progress with Japan.

The next day I told the Prime Minister of Great Britain and Generalissimo Stalin that the explosion had been a success. The British Prime Minister understood and appreciated what I'd told him. Premier Stalin smiled and thanked me for reporting the explosion to him, but I'm sure he did not understand its significance.

I called a meeting of the Secretary of State, Mr. Byrnes, the Secretary of War, Mr. Stimson, Admiral Leahy, General Marshall, General Eisenhower, Admiral King and some others, to discuss what should be done with this awful weapon.

I asked General Marshall what it would cost in lives to land on the Tokio plain and other places in Japan. It was his opinion that such an invasion would cost at a minimum one quarter of a million casualties, and might cost as much as a million, on the American side alone, with an equal number of the enemy. The other military and naval men present agreed.

I asked Secretary Stimson which cities in Japan were devoted exclusively to war production. He promptly named Hiroshima and Nagasaki, among others.

We sent an ultimatum to Japan. It was rejected.

I ordered atomic bombs dropped on the two cities named on the way back from Potsdam, when we were in the middle of the Atlantic Ocean. . . .

Dropping the bombs ended the war, saved lives, and gave the free nations a chance to face the facts.

2. Among those consulted on the decision to use the atomic bomb, none was more universally respected than British Prime Minister Winston Churchill. The discussions he describes in the following excerpt from his book *Triumph and Tragedy* **took place during the Potsdam Conference:**[15]

On July 17 world-shaking news had arrived. In the afternoon Stimson called at my abode and laid before me a sheet of paper on which was written, "Babies satisfactorily born." By his manner I saw something extraordinary had happened. "It means," he said, "that the experiment in the New Mexican desert has come off. The atomic bomb is a reality." Although we had followed this dire quest with every scrap of information imparted to us, we had not been told beforehand, or at any rate I did not know, the date of the decisive trial. No responsible scientist would predict what would happen when the first full-scale atomic explosion was tried. Were these bombs useless or were they annihilating? Now we knew. . . .

Here then was a speedy end to the Second World War, and perhaps to much else besides.

The President invited me to confer with him forthwith. He had with him General Marshall and Admiral Leahy. Up to this moment we had shaped our ideas towards an assault upon the homeland of Japan by terrific air bombing and by the invasion of very large armies. We had contemplated the desperate resistance of the Japanese fighting to the death with Samurai devotion, not only in pitched battles, but in every cave and dug-out. I had

---

[15] Winston S. Churchill, *Triumph and Tragedy* (Houghton Mifflin Co., and Cassell and Co., Ltd., Copyright 1953), pp. 637-639.

in my mind the spectacle of Okinawa island, where many thousands of Japanese, rather than surrender, had drawn up in line and destroyed themselves by hand-grenades after their leaders had solemnly performed the rite of *hara-kiri*. To quell the Japanese resistance man by man and conquer the country yard by yard might well require the loss of a million American lives and half that number of British—or more if we could get them there: for we were resolved to share the agony. Now all this nightmare picture had vanished. In its place was the vision—fair and bright indeed it seemed—of the end of the whole war in one or two violent shocks. I thought immediately myself of how the Japanese people, whose courage I had always admired, might find in the apparition of this almost supernatural weapon an excuse which would save their honour and release them from their obligation of being killed to the last fighting man....

At any rate, there never was a moment's discussion as to whether the atomic bomb should be used or not. To avert a vast, indefinite butchery, to bring the war to an end, to give peace to the world, to lay healing hands upon its tortured peoples by a manifestation of overwhelming power at the cost of a few explosions, seemed, after all our toils and perils, a miracle of deliverance.

British consent in principle to the use of the weapon had been given on July 4, before the test had taken place. The final decision now lay in the main with President Truman, who had the weapon; but I never doubted what it would be, nor have I ever doubted since that he was right. The historic fact remains, and must be judged in the after-time, that the decision whether or not to use the atomic bomb to compel the surrender of Japan was never even an issue. There was unanimous, automatic, unquestioned agreement around our table; nor did I ever hear the slightest suggestion that we should do otherwise.

E. EVIDENCE AFTER THE FACT

How accurate were Allied estimates of Japan's ability to continue the war and resist an invasion? The evidence presented below was not available when the decision to bomb was made. It was gathered by Allied investigators who entered Japan immediately after the surrender as well as from other sources not available until after the war.

1. Under interrogation by American officers after the war, two top-ranking officers of the Japanese air force explained their plan for meeting the expected invasion of Japan:[16]

Lieutenant General Tazoe[17] ...

"The air force plan was to attack the Allied fleet by Kamikaze planes, and for that purpose the full air force led by the commanding general was made ready to destroy the Allied ships near the shore. We expected annihilation of our entire air force, but we felt that it was our duty. The army and navy each had 4,000-5,000 planes for this purpose. Of that force, waves of 300-400 planes at the rate of one wave per hour for each of the army and navy would have been used to oppose a landing on Kyushu.

---

[16] Assistant Chief of Air Staff-Intelligence, Headquarters, Army Air Forces, *Mission Accomplished: Interrogations of Japanese Industrial, Military and Civil Leaders of World War II* (Washington, Government Printing Office, 1946), pp. 34-35.

[17] Lt. Gen. Noburu Tazoe was Chief of Staff, Air General Army, the equivalent of the U.S. Army Air Force.

"We thought we could win the war by using Kamikaze planes on the ships offshore; the ground forces would handle those which got through. The army could not put out effective resistance without the air arm, but we intended doing the best we could even if we perished. The entire navy and army air forces volunteered as Kamikaze and there was sufficient fuel for these attacks.

"Based on the Leyte and Okinawa experiences, it was contemplated that one out of four planes (of the 8,000-9,000 available for special attack) would sink or damage an Allied ship....

"The air general army had been following a policy of conserving aircraft for the purpose of countering the expected invasion....We had 5,000 pilots with enough experience for special attack against invasion and 3,000 more in training...."

General Kawabe[18] ...

"I know that you in the United States found it more difficult to manufacture crews than planes and did everything possible to rescue the crews, but our strategy was aimed solely at the destruction of your fleet and transport fleet when it landed in Japan. It was not very difficult to manufacture second-rate planes, that is, makeshift planes, and it was not difficult to train pilots for just such a duty; and since pilots were willing, we had no shortage of volunteers....

"But, I wish to explain something, which is a very difficult thing and which you may not be able to understand. The Japanese, to the very end, believed that by spiritual means they could fight on equal terms with you, yet by any other comparison it would not appear equal. We believed our spiritual confidence in victory would balance any scientific advantages and we had no intention of giving up the fight....

"You call our Kamikaze attacks suicide attacks. This is a misnomer and we feel very badly about your calling them suicide attacks. They were in no sense suicide. The pilot did not start out on his mission with the intention of committing suicide. He looked upon himself as a human bomb which would destroy a certain part of the enemy fleet for his country. They considered it a glorious thing, while suicide may not be so glorious."

2. A Kamikaze pilot writes his last letter home:[19]

Do not weep because I am about to die. If I were to live and one of my dear ones to die, I would do all in my power to cheer those who remain behind. I would try to be brave.

11:30 a.m.—the last morning. I shall now have breakfast and then go to the aerodrome. I am busy with my final briefing and have no time to write any more. So I bid you farewell.

Excuse this illegible letter and the jerky sentences.

Keep in good health.

I believe in the victory of Greater Asia.

I pray for the happiness of you all, and I beg your forgiveness for my lack of piety.

---

[18]Lt. Gen. Masakazu Kawabe was Commanding General, Air General Army and Director of Kamikaze Operations, Philippines and Okinawa Campaigns.

[19]Desmond Flower and James Reeves, eds., *The Taste of Courage: The War, 1939-1945* (Tokyo, Japan, University of Tokyo Press, 1960). p. 743.

I leave for the attack with a smile on my face. The moon will be full to-night. As I fly over the open sea off Okinawa I will choose the enemy ship that is to be my target.

I will show you that I know how to die bravely.

With all my respectful affection,

Akio Otsuka

3. In April, 1945, the aged Admiral Baron Kantaro Suzuki was appointed Prime Minister of Japan, with explicit instructions from the Emperor to find some honorable way to end the war. One of Suzuki's first actions was to order his chief cabinet secretary to carry out a detailed survey of the Japanese resources as they stood at the time. Here are the highlights of his secretary's confidential report, depicting the situation in early June, 1945:[20]

A. General

The ominous turn of the war, coupled with the increasing tempo of air raids is bringing about great disruption of land and sea communications and essential war production. The food situation has worsened. It has become increasingly difficult to meet the requirements of total war. Moreover, it has become necessary to pay careful attention to the trends in public sentiment.

B. National Trends in General

Morale is high, but there is dissatisfaction with the present regime. Criticisms of the government and the military are increasing. The people are losing confidence in their leaders, and the gloomy omen of deterioration of public morale is present. The spirit of public sacrifice is lagging and among leading intellectuals there are some who advocate peace negotiations as a way out. . . .

C. Manpower

1. As compared with material resources, there is a relative surplus of manpower, but there is no efficient exploitation of it. . . .

2. The physical standard and birth rate of the people are on the down grade. . . .

D. Transportation and Communication

Transportation is faced with insurmountable difficulties because of fuel shortages, mounting fury of enemy attacks on our lines of communications, and insufficient manpower in cargo handling. . . .

Transport capacity of the railways will drop to half that of the previous year due to the enemy air attack and our inability to maintain construction and repairs on an efficient level. It is feared that railway transportation will become confined to local areas. . . .

E. Material Resources . . .

There is a strong possibility that a considerable portion of the various industrial areas will have to suspend operation for lack of coal. . . .

With oil reserves on the verge of exhaustion and the delay in plans for increased output of oil, we are faced with an extreme shortage of aviation fuel. . . .

It is becoming increasingly difficult to maintain production of aircraft. . . .

---

[20]Hisatsune Sakomizu, "Survey of Natural Resources as of 1-10 June 1945," included as Appendix A-2 in The United States Strategic Bombing Survey, *Japan's Struggle to End the War* (Washington, Government Printing Office, 1946), pp. 16-18.

F. National Living Conditions

1. *Foodstuffs.* The food situation has grown worse and a crisis will be reached at the end of this year. The people will have to get along on an absolute minimum of rice and salt required for subsistence considering the severity of air raids, difficulties in transportation, and the appearance of starvation conditions in the isolated sections of the nation. . . .

2. *Living conditions.* From now on prices will rise sharply bringing on inflation which will seriously undermine the wartime economy. . . .

4. The author of the following selection is a native-born, white American woman from Tennessee who, in 1931, met, fell in love with, and married a young Japanese diplomat. She lived through the war years in Japan with her husband and young daughter. After the war she described her experiences in a revealing book, from which are drawn the three brief passages presented below. They portray incidents that took place in 1944 and early 1945:[21]

Only victories were broadcast. This involved such obvious contradictions that even the more simple-minded listeners became doubtful. Everyone who could think at all realized that the country was in a more and more desperate state, its back to the wall. . . . The conviction of ultimate defeat had become widespread but everyone was careful not to speak his opinion. . . .

By many little signs I knew how desperate things had become for the Japanese. I saw little boys of ten and twelve unloading the freight from the trains. Children were employed in all kinds of factory work from clothes-making to riveting airplane parts together; they were mobilized through their schools and taken from there to their jobs each day by the teachers. . . .

A man had come out of the mist in front of them and was moving toward a huge trash can, placed at the corner of the sea wall near the steps, which was filled with garbage and refuse. Before the war there had been no beggars in Japan. The little girl and her father watched. The man grabbed into the barrel like an animal, spilling litter out onto the clean beach. He found some food clinging to a paper wrapper and he pressed the paper tightly against his face to gnaw the food away. . . .

Our house was to be the meeting place of the next *tonarigumi* [a unit of local government] gathering. There were to be new instructions from the *chokai* (town assembly). Our neighbors were farmers, fishermen, and a few families from Tokyo who had sought refuge there. . . .

[Every] person of adult age must provide himself with a bamboo spear of a certain length with which to meet the enemy when they came to invade the islands. I was so shocked by this that I sat in stunned silence. . . . They began to argue over the length of the spears and I was engulfed by waves of talk, most of which I could not understand. I thought of how these poor people would feel when they discovered how reckless their leaders had been. With docility and courage they were doing all in their power to stave off the inevitable, but it would come and they would know that they never had the least chance to win. . . . Then they filed out, after much bowing. I knew they were still wondering where, with their bamboo spears, they would go when the marines landed in our midst.

---

[21] Gwen Terasaki, *Bridge to the Sun* (Chapel Hill, University of North Carolina Press, 1957), pp. 134, 135, 148-150.

5. The United States Strategic Bombing Survey was established by Presidential order toward the end of the war to investigate the effects of our aerial attacks on Germany and Japan. The Survey's inquiries in Japan included interrogations of more than 700 leading figures of the Japanese government, armed forces and industry. The Survey also recovered and translated many documents, such as the Sakomizu report presented above. The Survey's own final conclusions as to the various factors that contributed to Japan's surrender were published in 1946. Excerpts follow:[22]

1. Blockade of Japan's sea communications exploited the basic vulnerability of an island enemy which, with inherently second-power resources, was struggling to enlarge its capabilities by milking the raw materials of a rich conquered area.... The blockade prevented exploitation of conquered resources, kept Japan's economy off balance, created shortages of materials which in turn limited war production, and deprived her of oil in amounts sufficient to immobilize fleet and air units and to impair training.... The direct military and economic limitations imposed by shortages created virtually insoluble political as well as economic problems.... The special feeling of vulnerability to blockade, to which a dependent island people are ever subject, increased and dramatized, especially to the leaders, the hopelessness of their position and favored the growing conviction that the defeat was inevitable.

2. While the blockade was definitive in strangling Japan's war mobilization and production, it cannot be considered separately from the pressure of our concurrent military operations, with which it formed a shears that scissored Japan's military potential into an ineffectual remnant. In the early engagements that stemmed the Japanese advance and in the subsequent battle for bases, the application of our air power ... enabled us largely to destroy her navy and reduce her air forces to impotence before the home islands could be brought under direct air attack.... Japan's principal land armies were in fact never defeated, a consideration which also supported the [Japanese] military's continued last-ditch resistance to the surrender decision. It nevertheless appears that after the loss of the Marianas in July-August 1944, the military commands, though unconvinced of final victory, viewed defense against our subsequent operations as affording an opportunity for only a limited success, a tactical victory which might, so they hoped, have created a purchase from which to try for a negotiated peace under terms more favorable than unconditional surrender.

3. ... The timing of the strategic bombing attack affected its role in the surrender decision. After the Marianas were lost but before the first attacks were flown in November 1944, Tojo had been unseated and peacemakers introduced into the Government as prominent elements.... These attacks became definitive in the surrender decision because they broadened the realization of defeat by bringing it home to the people and dramatized to the whole nation what the small peace party already knew. They proved day in and day out, and night after night, that the United States did control the air and could exploit it....

4. When Japan was defeated without invasion, a recurrent question arose as to what effect the threat of a home-island invasion had had upon the surrender decision. It was contended that the threat of invasion, if not the

---

[22]United States Strategic Bombing Survey, *Japan's Struggle to End the War* (Washington, Government Printing Office, 1946), pp. 10-13.

actual operation, was a requirement to induce acceptance of the surrender terms. On this tangled issue the evidence and hindsight are clear. The fact is, of course, that Japan did surrender without invasion, and with its principal armies intact.... The responsible leaders in power read correctly the true situation and embraced surrender well before invasion was expected.

5. So long as Germany remained in the war that fact contributed to the core of Japanese resistance.... The significant fact, however, is that Japan was pursuing peace before the Nazis collapsed, and the impoverishment and fragmentation of the German people had already afforded a portent of similar consequences for an intransigent Japan.

6. The Hiroshima and Nagasaki atomic bombs did not defeat Japan, nor by the testimony of the enemy leaders who ended the war did they persuade Japan to accept unconditional surrender. The Emperor, the Lord Privy Seal, the Prime Minister, the Foreign Minister, and the Navy Minister had decided as early as May of 1945 that the war should be ended even if it meant acceptance of defeat on allied terms. The War Minister and the two chiefs of staff opposed unconditional surrender. The impact of the Hiroshima attack was to bring further urgency and lubrication to the machinery of achieving peace, primarily by contributing to a situation which permitted the Prime Minister to bring the Emperor overtly and directly into a position where his decision for immediate acceptance of the Potsdam Declaration could be used to override the remaining objectors. Thus, although the atomic bombs changed no votes of the Supreme War Direction Council concerning the Potsdam terms, they did foreshorten the war and expedite the peace....

Based on a detailed investigation of all the facts and supported by the testimony of the surviving Japanese leaders involved, it is the Survey's opinion that certainly prior to 31 December 1945, and in all probability prior to 1 November 1945, Japan would have surrendered even if the atomic bombs had not been dropped, even if Russia had not entered the war, and even if no invasion had been planned or contemplated.

SECTION 3

The Agony of the Atomic Scientists

As the time drew near for the final showdown, two proposals were made for inducing the Japanese to surrender short of an invasion or the dropping of the bomb. One of these proposals will be considered in this section. It was advocated by some, but not all, of the atomic scientists and won the support of some important political figures as well. The other proposal will be discussed in Section 4. Whether one idea or both might have worked will forever remain one of history's more tantalizing questions.

No more fearsome specter haunted the Allied statesmen of World War II than the possibility that Germany might be first to produce an atomic weapon. Germany had led the world in nuclear physics since the turn of the century. After the advent of Hitler in 1933, there were indications that the Nazi regime was sponsoring a secret atomic research project.

Among those most concerned were a number of top European scientists who had been driven into exile by the rise of the Nazis. Some of these had been dismissed from German universities and laboratories because they were Jewish and hence anathema to the anti-Semitic Nazis. Others resigned and left Germany voluntarily rather than serve the Nazi regime. The pattern was repeated elsewhere in western Europe as the Nazi drive for world conquest gained ground in the late 1930's and early 40's. Even with their departure enough first-class scientists remained in Germany to keep the danger of a Nazi A-bomb alive.

In the United States the refugee scientists worked effectively with their American colleagues in what many saw as a race to develop the bomb first. The brilliant refugee roster included among others: Niels Bohr of Denmark; Albert Einstein, James Franck, Eugene Wigner, and Hans Bethe of Germany; Leo Szilard and Edward Teller of Hungary; and Enrico Fermi of Italy. The historic first step was a letter, drafted by Leo Szilard and signed by Albert Einstein, addressed to President Franklin D. Roosevelt on August 2, 1939. It informed the President that recent discoveries indicated that a new form of energy might now be derived from atomic fission, that this atomic energy might be usable in an incredibly powerful new weapon, that the Germans might already be developing such a weapon, and that the U. S. government should therefore take steps to encourage atomic research. Out of this letter eventually came the top-secret, two-billion-dollar "Manhattan Project" which, nearly six years later, produced the bomb.

Then, early in 1945, with the bomb still several months from completion, startling news reached the scientists. Special teams of investigators, racing into Germany with advance units of Allied troops, discovered that the Germans were not even close to developing an atomic bomb. The discovery produced a grave crisis of conscience for some of those very scientists who had contributed most to creating the bomb. With Germany out of the war and Japan visibly weakening, could not some alternative be found to prevent their monstrous brainchild from being unleashed upon the world?

1. The atomic scientists discussed this problem at all of the bomb project's far-flung installations, but the chief center of agitation was the Metallurgical Laboratory at the University of Chicago. Here, on June 11, 1945, seven scientists of the Committee of Social and Political Implications, headed by Nobel Prize winner James Franck, produced what has become the most celebrated protest statement of the period. Noteworthy among its signers was Leo Szilard, the man who, with Einstein, had fathered the whole project. Franck went to Washington and tried to present the petition personally to Secretary of War Stimson, but Stimson was out of town and the petition had to be left with one of his assistants.[1]

> The scientists on this Project do not presume to speak authoritatively on problems of national and international policy. However, we found ouselves, by the force of events, during the last five years, in the position of a small group of citizens cognizant of a grave danger for the safety of this country as well as for the future of all the other nations, of which the rest of mankind is unaware.... We believe that our acquaintance with the scientific elements of the situation and prolonged pre-occupation with its world-wide political implications, imposes on us the obligation to offer ... some suggestions....
>
> One possible way to introduce nuclear weapons to one [sic] world—which may particularly appeal to those who consider nuclear bombs primarily as a secret weapon developed to help win the present war—is to use them without warning on appropriately selected objects in Japan....
>
> Russia, and even allied countries which bear less mistrust of our ways and intentions, as well as neutral countries may be deeply shocked by this step. It may be very difficult to persuade the world that a nation which was capable of secretly preparing and suddenly releasing a new weapon, as indiscriminate as the rocket bomb and a thousand times more destructive, is to be trusted in its proclaimed desire of having such weapons abolished by international agreement. We have large accumulations of poison gas, but do not use them, and recent polls have shown that public opinion in this country would disapprove of such a use even if it would accelerate the winning of the Far Eastern War.... [It] is not at all certain that American public opinion, if it could be enlightened as to the effect of atomic explosives, would approve of our own country being the first to introduce such an indiscriminate method of wholesale destruction of civilian life.
>
> Thus ... the military advantages and the saving of American lives achieved by the sudden use of atomic bombs against Japan may be outweighed by the ensuing loss of confidence and by a wave of horror and repulsion sweeping over the rest of the world and perhaps even dividing public opinion at home.
>
> From this point of view, a demonstration of the new weapon might best be made, before the eyes of representatives of all the United Nations, on the desert or a barren island. The best possible atmosphere for the achievement of an international agreement could be achieved if America could say to the world, "You see what sort of weapon we had but did not use. We are ready to renounce its use in the future if other nations join us in this renunciation and agree to the establishment of an efficient international control."
>
> After such a demonstration the weapon might perhaps be used against Japan if the sanction of the United Nations (and of public opinion at home)

---

[1] "Before Hiroshima: A Report to the Secretary of War, June 1945," *Bulletin of the Atomic Scientists* (May 1, 1946), pp. 2-4, 16.

were obtained, perhaps after a preliminary ultimatum to Japan to surrender or at least to evacuate certain regions as an alternative to their total destruction. This may sound fantastic, but in nuclear weapons we have something entirely new in order of magnitude of destructive power....

Nuclear bombs cannot possibly remain a "secret weapon" at the exclusive disposal of this country for more than a few years. The scientific facts on which their construction is based are well known to scientists of other countries. Unless an effective international control of nuclear explosives is instituted, a race for nuclear armaments is certain to ensue following the first revelation of our possession of nuclear weapons to the world. Within ten years other countries may have nuclear bombs, each of which weighing less than a ton, could destroy an urban area of more than ten square miles. In the war to which such an armaments race is likely to lead, the United States, with its agglomeration of population and industry in comparatively few metropolitan districts, will be at a disadvantage compared to nations whose population and industry are scattered over large areas.

2. Perhaps the most brilliant galaxy of scientific talent ever assembled was that which worked at the super-secret atomic bomb laboratory high atop a mesa at remote Los Alamos, New Mexico. Here, too, the scientists debated how and whether the bomb should be used, but no concerted action was ever taken. Edward Teller, a key physicist at Los Alamos and later famous as "father of the H-bomb," indicated at least part of the reason in his recollections published in 1962:[2]

In the spring of 1945, I did become worried about the way the atomic bomb might be used. My apprehension reached a high plateau several months before Hiroshima when I received a letter at Los Alamos from Szilard. He asked my support for a petition urging that the United States would not use the atomic bomb in warfare without first warning the enemy.

I was in absolute agreement, and prepared to circulate Szilard's petition among the scientists at Los Alamos. But it was my duty, first, to discuss the question with the director of the Los Alamos Laboratory, Dr. J. Robert Oppenheimer. He was the constituted authority at Los Alamos. But he was more: His brilliant mind, his quick intellect, and his penetrating interest in everyone at the laboratory made him our natural leader as well. He seemed to be the obvious man to turn to with any formidable problem, particularly political.

Oppenheimer told me, in a polite and convincing way, that he thought it improper for a scientist to use his prestige as a platform for political pronouncements. He conveyed to me in glowing terms the deep concern, thoroughness, and wisdom with which these questions were being handled in Washington. Our fate was in the hands of the best, the most conscientious men of our nation. And they had information which we did not possess. Oppenheimer's words lifted a great weight from my heart. I was happy to accept his word and his authority. I did not circulate Szilard's petition. Today I regret that I did not....

We could have exploded the bomb at a very high altitude over Tokyo in the evening. Triggered at a high altitude, the bomb would have created a sudden, frightening daylight over the city. But it would have killed no one. After the bomb had been demonstrated— after we were sure that it was not

---

[2]Edward Teller, with Allen Brown, *The Legacy of Hiroshima* (New York, Doubleday, 1962), pp. 13-14.

a dud—we could have told the Japanese what it was and what would happen if another atomic bomb were detonated at low altitude. . . .

The ultimatum, I believe, would have been met, and the atomic bomb could have been used more humanely but just as effectively to bring a quick end to the war. But, to my knowledge, such an unannounced, high-altitude demonstration over Tokyo at night was never proposed.

3. In May 1945 President Truman appointed a high-level "Interim Committee," with Stimson as chairman, to advise him on all implications of atomic energy. A distinguished Scientific Panel of atomic specialists was set up as consultant to the Committee: Arthur H. Compton, J. Robert Oppenheimer, Enrico Fermi, Ernest O. Lawrence. The Interim Committee and Scientific Panel held their first joint meeting on May 31, 1945.

In the following excerpt from his memoirs Arthur Compton, a member of the Scientific Panel, recalls the discussions on that occasion and then goes on to describe the ensuing controversies that split wide open the tight little community of atomic scientists.[3]

Throughout the morning's discussions it seemed to be a foregone conclusion that the bomb would be used. It was regarding only the details of strategy and tactics that differing views were expressed. At the luncheon following the morning meeting, I was seated at Mr. Stimson's left. In the course of the conversation I asked the Secretary whether it might not be possible to arrange a nonmilitary demonstration of the bomb in such a manner that the Japanese would be so impressed that they would see the uselessness of continuing the war. The Secretary opened this question for general discussion by those at the table. Various possibilities were brought forward. One after the other it seemed necessary that they should be discarded.

It was evident that everyone would suspect trickery. If a bomb were exploded in Japan with previous notice, the Japanese air power was still adequate to give serious interference. An atomic bomb was an intricate device, still in the developmental stage. Its operation would be far from routine. If during the final adjustments of the bomb the Japanese defenders should attack, a faulty move might easily result in some kind of failure. Such an end to an advertised demonstration of power would be much worse than if the attempt had not been made. It was now evident that when the time came for the bombs to be used we should have only one of them available, followed afterwards by others at all-too-long intervals. We could not afford the chance that one of them might be a dud. If the test were made on some neutral territory, it was hard to believe that Japan's determined and fanatical military men would be impressed. If such an open test were made first and failed to bring surrender, the chance would be gone to give the shock of surprise that proved so effective. On the contrary, it would make the Japanese ready to interfere with an atomic attack if they could. Though the possibility of a demonstration that would not destroy human lives was attractive, no one could suggest a way in which it could be made so convincing that it would be likely to stop the war.

After luncheon the Interim Committee went into executive session. Our Scientific Panel was then again invited in. We were asked to prepare a report

---
[3]Arthur H. Compton, *Atomic Quest* (New York, Oxford University Press, 1956), pp. 238-241, 242-244, 246-247.

as to whether we could devise any kind of demonstration that would seem likely to bring the war to an end without using the bomb against a live target.

Ten days later, at Oppenheimer's invitation, Lawrence, Fermi, and I spent a long week end at Los Alamos. We were keenly aware of our responsibility as the scientific advisers to the Interim Committee. Among our colleagues were the scientists who supported Franck in suggesting a nonmilitary demonstration only. We thought of the fighting men who were set for an invasion which would be so very costly in both American and Japanese lives. We were determined to find, if we could, some effective way of demonstrating the power of an atomic bomb without loss of life that would impress Japan's warlords. If only this could be done!

Ernest Lawrence was the last one of our group to give up hope for finding such a solution. The difficulties of making a purely technical demonstration that would carry its impact effectively into Japan's controlling councils were indeed great. We had to count on every possible effort to distort even obvious facts. Experience with the determination of Japan's fighting men made it evident that the war would not be stopped unless these men themselves were convinced of its futility. Secretary Stimson has published the following paragraph which constituted the essence of our report:

> The opinions of our scientific colleagues on the initial use of these weapons are not unanimous: they range from the proposal of a purely technical demonstration to that of the military application best designed to induce surrender. Those who advocate a purely technical demonstration would wish to outlaw the use of atomic weapons, and have feared that if we use the weapons now our position in future negotiations will be prejudiced. Others emphasize the opportunity of saving American lives by immediate military use, and believe that such use will improve the international prospects, in that they are more concerned with the prevention of war than with the elimination of this special weapon. We find ourselves closer to these latter views; [we can propose no technical demonstration likely to bring an end to the war, we see no acceptable alternative to direct military use.] (The brackets are Mr. Stimson's.)[4]

Our hearts were heavy as on 16 June we turned in this report to the Interim Committee. We were glad and proud to have had a part in making the power of the atom available for the use of man. What a tragedy it was that this power should become available first in time of war and that it must first be used for human destruction. If, however, it would result in the shortening of the war and the saving of lives—if it would mean bringing us closer to the time when war would be abandoned as a means of settling international disputes—here must be our hope and our basis for courage....

The statement of our Scientific Panel to the Interim Committee was in close accord with the views of most of the scientists engaged on the atomic project. The case against use of the bomb in the Japanese theater was pressed most vigorously by Leo Szilard. Fearful lest more routine procedures be ineffective, he wrote a letter direct to the President and went to Washington to urge personally that the use of the bomb be blocked. He

---

[4]Henry L. Stimson, "The Decision to Use the Atomic Bomb," pp. 101.

circulated petitions at Chicago and urged others to circulate similar petitions at Los Alamos and Oak Ridge requesting that the atomic bomb should not be used in World War II. This action stimulated counterpetitions requesting that as soon as the bombs were available they should be used as might be necessary to bring the war to a close. . . .

One counterpetition read in part thus:

> Are not the men of the fighting forces a part of the nation? Are not they, who are risking their lives for the nation entitled to the weapons which have been designed? In short, are we to go on shedding American blood when we have available a means to speedy victory? No! If we can save even a handful of American lives, then let us use this weapon—now!
> ... These sentiments, we feel, represent more truly those of the majority of Americans and particularly those who have sons ... In the foxholes and warships in the Pacific.

One of the young men who had been with us at Chicago and had transferred to Los Alamos came into my Chicago office in a state of emotional stress. He said he had heard of an effort to prevent the use of the bomb. Two years earlier I had persuaded this young man, as he was graduating with a major in physics, to cast his lot with our project. The chances are, I had told him, that you will be able to contribute more toward winning the war in this position than if you should accept the call to the Navy that you are considering. He had heeded my advice. Now he was sorely troubled: 'I have buddies who have fought through the battle of Iwo Jima. Some of them have been killed, others wounded. We've got to give these men the best weapons we can produce.' Tears came to his eyes. 'If one of these men should be killed because we didn't let them use the bombs, I would have failed them. I just could not make myself feel that I had done my part.' Others, though less emotional, felt just as deeply.

An especially carefully considered petition that carried a large number of signatures read as if those who framed it had been reading the minds of Mr. Truman and Mr. Stimson. It was addressed to the President:

> ... We respectfully petition that the use of atomic bombs, particularly against cities, be sanctioned by you as Chief Executive only under the following conditions:
>
> 1. Opportunity has been given to the Japanese to surrender on terms assuring them the possibility of peaceful development in their homeland.
>
> 2. Convincing warnings have been given that refusal to surrender will be followed by the use of a new weapon.
>
> 3. Responsibility for the use of atomic bombs is shared with our allies.

It was difficult from such petitions to get a balanced view of how our men were thinking. General Groves accordingly suggested that I supervise an opinion poll among those who knew what was going on. Farrington Daniels, then Director of the Metallurgical Laboratory, took charge of the poll at Chicago. Oppenheimer at Los Alamos and Lawrence at Berkeley used less formal methods of sounding the opinions of their men. It was attitudes thus expressed that the Scientific Panel had in mind as we wrote our report to the Interim Committee reluctantly rejecting the nonmilitary demonstration of the bomb.

Daniels asked and received replies from 150 members of the Metallurgical Laboratory at Chicago. His questionnaire had five procedures, graded from no use of the bomb in this war to its military use in the manner most effective in bringing prompt Japanese surrender. There were a few who preferred not to use the bomb at all, but 87 per cent voted for its military use, at least if after other means were tried this was found necessary to bring surrender. . . .

[On 23 July] Colonel Nichols came to me at Oak Ridge with the word, 'Washington wants at once the results of the opinion polls on the use of the bomb.' I knew of the conference at Potsdam, but of course knew nothing of the state of its discussions. The votes and petitions were by now in my hands. I accordingly wrote out a message summarizing the results as objectively as I could and handed it to the Colonel. An hour later he came to me again. 'Washington wants to know what you think.'

What a question to answer! Having been in the very midst of these discussions, it seemed to me that a firm negative stand on my part might still prevent an atomic attack on Japan. Thoughts of my pacifist Mennonite ancestors flashed through my mind. I knew all too well the destruction and human agony the bombs would cause. I knew the danger they held in the hands of some future tyrant. These facts I had been living with for four years. But I wanted the war to end. I wanted life to become normal again. I saw a chance for an enduring peace that would be demanded by the very destructiveness of these weapons. I hoped that by use of the bombs many fine young men I knew might be released at once from the demands of war and thus be given a chance to live and not to die.

'My vote is with the majority. It seems to me that as the war stands the bomb should be used, but no more drastically than needed to bring surrender.'

Colonel Nichols took my message and sent it at once to Washington. Two weeks later the first bomb fell on Hiroshima.

4. On June 27 Undersecretary of the Navy Ralph A. Bard delivered the following memorandum to the Interim Committee, thereby becoming the only member of that Committee to dissent formally from its recommendation. The text of the memorandum was published fifteen years later in a news magazine's analysis of the decision:[5]

Ever since I have been in touch with this program, I have had a feeling that before the bomb is actually used against Japan that Japan should have some preliminary warning for say two or three days in advance of use. The position of the United States as a great humanitarian nation and the fair play attitude of our people generally is responsible in the main for this feeling.

During recent weeks I have also had the feeling very definitely that the Japanese government may be searching for some opportunity which they could use as a medium for surrender. Following the three-power conference [at Potsdam] emissaries from this country could contact representatives

---

[5]"President Truman Did Not Understand," *U. S. News and World Report,* August 15, 1960. p. 74.

from Japan somewhere on the China coast and make representations with regard to Russia's position and at the same time give them some information regarding the proposed use of atomic power, together with whatever assurances the President might care to make with regard to the Emperor of Japan and the treatment of the Japanese nation following unconditional surrender. It seems quite possible to me that this presents the opportunity which the Japanese are looking for.

I don't see that we have anything in particular to lose in following such a program. The stakes are so tremendous that it is my opinion very real consideration should be given to some plan of this kind. I do not believe under present circumstances existing that there is anyone in this country whose evaluation of the chances of success of such a program is worth a great deal. The only way to find out is to try it out.

5. In his memoirs Secretary of State Henry L. Stimson recounted his and the Interim Committee's recommendations to the President:[6]

The Interim Committee, on June 1, recommended that the bomb should be used against Japan, without specific warning, as soon as possible, and against such a target as to make clear its devastating strength. Any other course, in the opinion of the committee, involved serious danger to the major objective of obtaining a prompt surrender from the Japanese. An advisory panel of distinguished atomic physicists reported that "We can propose no technical demonstration likely to bring an end to the war; we see no acceptable alternative to direct military use."

"The committee's function was, of course, entirely advisory. The ultimate responsibility for the recommendation to the President rested upon me, and I have no desire to veil it. The conclusions of the committee were similar to my own, although I reached mine independently. I felt that to extract a genuine surrender from the Emperor and his military advisers, there must be administered a tremendous shock which would carry convincing proof of our power to destroy the Empire. Such an effective shock would save many times the number of lives, both American and Japanese, that it would cost.

6. The final decision devolved upon President Truman. The following statement was made by Truman to his biographer, William Hillman:[7]

"We were planning an invasion of Japan with the use of 2,000,000 men and the military had estimated the invasion might result in very heavy casualties. In April I had appointed an interim committee to make recommendations on questions of policy when and if an atomic bomb could be made. The Committee consisted of Secretary Stimson, George L. Harrison, James H. Byrnes, William L. Clayton, Dr. Vannevar Bush, Dr. Karl T. Compton, and Dr. James B. Conant. Before I had left for Potsdam the committee had recommended that the bomb be used against Japan.

---

[6] Henry L. Stimson and McGeorge Bundy, *On Active Service in Peace and War* (Harper & Row and Hutchinson Pub. Group Ltd. and Curtis Brown Ltd.), p. 617, Copyright 1947, 1948 by H. L. Stimson.
[7] William Hillman, *Mr. President* (New York, Farrar, Straus and Giroux, copyright 1952), pp. 248-249.

—"General Marshall said in Potsdam that if the bomb worked we would save a quarter of a million American lives and probably save millions of Japanese.

"I gave careful thought to what my advisers had counseled. I wanted to weigh all the possibilities and implications. Here was the most powerful weapon of destruction ever devised and perhaps it was more than that. . . .

"I then agreed to the use of the atomic bomb if Japan did not yield.

"I had reached a decision after long and careful thought. It was not an easy decision to make. I did not like the weapon. But I had no qualms if in the long run millions of lives could be saved.

"The rest is history."

SECTION 4

Was It the Product of a Diplomatic Blunder— or of Diplomatic Calculation?

Subtle and extremely complicated questions of international relations played a part in making the decision to use the atomic bomb. Negotiations with the enemy, and relations with our wartime allies, were both involved.

A. THE "UNCONDITIONAL SURRENDER" PROBLEM

By the spring of 1945 the more realistic among Japan's leaders had recognized that their military fortunes were worsening rapidly. They began to cast about for some means to end the war but still hoped to avoid total humiliation by obtaining a negotiated peace. This plan faced one great obstacle: the insistence by the United States and her allies on the "unconditional surrender" of the enemy as the only peace offer they would consider.

The idea of "unconditional surrender" was neither willful nor accidental. While in part it resulted from resentment of Japan's sneak attack on Pearl Harbor, in larger part it was a direct response to what seemed to be the basic causes of the war. For at least in Europe, World War II was essentially a recurrence of World War I, which had ended scarcely more than twenty years before the second war broke out. The first war had not been fought through to unconditional surrender. The German army had been beaten but not destroyed. The peace settlement that followed was at least in some senses a negotiated peace, and the Nazis rose to power in the thirties partly on the argument that it was an unfair peace and should never have been accepted in the first place. As the peace settlement broke down and war broke out again, Americans and their allies vowed that they would not make the same mistake a second time, and that the only way to assure a lasting peace was to fight through to total victory, forcing the enemy to surrender without conditions or negotiations of any kind. "Unconditional surrender" was thus from the beginning the American and Allied goal of the war.

1. Early in July, after some desultory Japanese peace feelers through Sweden and Switzerland had evoked no response, Japan decided to request the Soviet Union to mediate with the Allies. At that time Russia had not yet entered the war against Japan. None of Japan's leaders knew that the Russians, at American insistence, had already promised to enter the war.

The following are excerpts from the supposedly secret dispatches sent by Japanese Foreign Minister Togo to his ambassador in Moscow, dated July 11 and 12, 1945.[1] Many similar messages were sent in the ensuing weeks. The United States had broken

Japan's secret codes early in the war, and hence was able to intercept and decode all such messages. Throughout the weeks leading up to Hiroshima, President Truman and other American leaders were therefore aware of Japan's diplomatic maneuvers. No direct response was ever made.

> The foreign and domestic situation for the Empire is very serious, and even the termination of the war is now being considered privately. Therefore ... we are also sounding out the extent to which we might employ the U.S.S.R. in connection with the termination of the war ... meet with Molotov immediately ... please explain our attitude as follows ...
>
> "We consider the maintenance of peace in Asia as one aspect of maintaining world peace. We have no intention of annexing or taking possession of the areas which we have been occupying as a result of the war....
>
> "His Majesty the Emperor is greatly concerned over the daily increasing calamities and sacrifices faced by the citizens of the various belligerent countries in this present war, and it is His Majesty's heart's desire to see the swift termination of the war. In the Greater East Asia War, however, as long as America and England insist on unconditional surrender, our country has no alternative but to see it through in an all-out effort for the sake of survival and the honor of the homeland...."

2. Few Americans thought as deeply about the demand for unconditional surrender and its probable effect on the Japanese as did Joseph C. Grew. He had been United States ambassador to Japan during the ten years prior to Pearl Harbor. Throughout the war he served as a key diplomatic advisor to Presidents Roosevelt and Truman. The following excerpt is from his *Memoirs*, published in 1952:[2]

> For a long time I had held the belief, based on my intimate experience with Japanese thinking and psychology over an extensive period, that the surrender of the Japanese would be highly unlikely, regardless of military defeat, in the absence of a public undertaking by the President that unconditional surrender would not mean the elimination of the present dynasty if the Japanese people desired its retention. I furthermore believed that if such a statement could be formulated and issued shortly after the great devastation of Tokyo by our B-29 attacks on or about May 26, 1945, the hands of the Emperor and his peace-minded advisers would be greatly strengthened in the face of the intransigent militarists and that the process leading to an early surrender might even then be set in motion by such a statement. Soviet Russia had not then entered the war against Japan, and since the United States had carried the major burden of the war in the Pacific, and since the President had already publicly declared that unconditional surrender would mean neither annihilation nor enslavement, I felt that the President would be fully justified in amplifying his previous statement as suggested. My belief in the potential effect of such a statement at that particular juncture was fully shared and supported by those officers in the Department of State who knew Japan and the Japanese well....

---

[1] *Potsdam Papers*, I, pp. 874-876.
[2] Joseph C. Grew, *Turbulent Era*, Walter Johnson, ed. (Boston, Houghton Mifflin, Copyright, 1952 by Joseph C. Grew), II, pp. 1421-1425.

Then, on my own initiative, as Acting Secretary of State, I called on President Truman on May 28, 1945, and presented this thesis as set forth in a memorandum prepared immediately after that meeting.... I also handed the President on May 28 a draft of a proposed statement which we in the State Department had prepared after long and most careful consideration.

In my own talk with the President on May 28, he immediately said that his own thinking ran along the same lines as mine, but he asked me to discuss the proposal with the Secretaries of War and Navy and the Chiefs of Staff.... A conference was therefore called and was held in the office of the Secretary of War in the Pentagon Building on May 29, 1945, and the issue was discussed for an hour. According to my memorandum of that meeting it became clear in the course of the discussion that Mr. Stimson, Mr. Forrestal [Secretary of the Navy], and General Marshall (Admiral King was absent) were all in accord with the principle of the proposal but that for certain military reasons, not then divulged, it was considered inadvisable for the President to make such a statement at that juncture. It later appeared that the fighting on Okinawa was still going on, and it was felt that such a declaration as I proposed would be interpreted by the Japanese as a confession of weakness. The question of timing was the nub of the whole matter, according to the views expressed. I duly reported this to the President, and the proposal for action was, for the time being, dropped....

Mr. Stimson did take energetic steps at Potsdam to secure the decision by the President and Mr. Churchill to issue the proclamation. In fact, the opinion was expressed to me by one American already in Potsdam, that if it had not been for Mr. Stimson's wholehearted initiative, the Potsdam Conference would have ended without any proclamation to Japan being issued at all. But even Mr. Stimson was unable to have included in the proclamation a categorical undertaking that unconditional surrender would not mean the elimination of the dynasty if the Japanese people desired its retention.

3. In his memoirs, published in 1947, Stimson commented on the Grew proposal:[3]

For months there had been disagreement at high levels over the proper policy toward the Emperor. Some maintained that the Emperor must go, along with all the other trappings of Japanese militarism. Others urged that the war could be ended much more cheaply by openly revising the formula of "unconditional surrender" to assure the Japanese that there was no intention of removing the Emperor.... This latter view had been urged with particular force and skill by Joseph C. Grew.... For their pains Grew and those who agreed with him were roundly abused as appeasers.

Stimson wholly agreed with Grew's general argument.... Unfortunately during the war years high American officials had made some fairly blunt and unpleasant remarks about the Emperor, and it did not seem wise to Mr. Truman and Secretary of State Byrnes that the Government should reverse its field too sharply; too many people were likely to cry shame....

The true question ... was not whether surrender could have been achieved without the use of the bomb but whether a different diplomatic and military course would have led to an earlier surrender.... But in the view

---

[3]Henry L. Stimson and McGeorge Bundy, *On Active Service In Peace and War*, pp. 626, 628-629.

of Stimson and his military advisers, it was always necessary to bear in mind that at least some of Japan's leaders would seize on any conciliatory offer as an indication of weakness. For this reason they did not support Grew in urging an immediate statement on the Emperor in May. The battle for Okinawa was proceeding slowly and with heavy losses, and they feared lest Japanese militarists argue that such a statement was the first proof of that American fatigue which they had been predicting since 1941. It seemed possible to Stimson, in 1947, that these fears had been based on a misreading of the situation....

Only on the question of the Emperor did Stimson take, in 1945, a conciliatory view; only on this question did he later believe that history might find that the United States, by its delay in stating its position, had prolonged the war.

4. On July 2, 1945, as they prepared to attend the summit conference at Potsdam, Stimson submitted to Truman a draft of a Declaration which it was proposed the leaders would issue at their meeting. The draft included the following as part of clause 12. It may be compared with the full text of the Declaration as it was finally approved and issued, which appears later in this section.[4]

The occupying forces of the Allies shall be withdrawn from Japan as soon as these objectives have been accomplished and there has been established beyond doubt a peacefully inclined, responsible government of a character representative of the Japanese people. This may include a constitutional monarchy under the present dynasty if the peace-loving nations can be convinced of the genuine determination of such a government to follow policies of peace which will render impossible the future development of aggressive militarism in Japan.

5. Opposition to this proposal was led by Assistant Secretary of State Archibald MacLeish, a distinguished scholar, poet, and former Librarian of Congress. On July 6 he addressed a memorandum to Secretary of State James F. Byrnes:[5]

What has made Japan dangerous in the past and will make her dangerous in the future if we permit it, is, in large part, the Japanese cult of emperor worship which gives the ruling groups in Japan—the *Gumbatsu* —the current coalition of militarists, industrialists, large land owners and office holders—their control over the Japanese people.... [The] institution of the throne is an anachronistic, feudal institution, perfectly adapted to the manipulation and use of anachronistic, feudal-minded groups within the country. To leave that institution intact is to run the grave risk that it will be used in the future as it has been used in the past. The argument most frequently advanced for the preservation of the throne is the argument that only the emperor can surrender. This is a powerful argument for the immediate future. It must be balanced against the longer-range consideration that however useful the emperor may be to us now, he may be a source of the greatest danger a generation from now. The same consideration applies to the argument that lives will be saved now if the Japanese are allowed to keep their emperor. The lives already spent will have been sacrificed in vain,

---
[4] *Potsdam Papers,* I, p. 899.
[5] *Ibid.,* pp. 896-897.

and lives will be lost again in the future in a new war, if the throne is employed in the future as it has been employed in the past by the Japanese Jingos and industrial expansionists.

6. On July 8 the Combined Intelligence Committee, a British-American group of military intelligence experts, reported to their superiors, the Combined Chiefs of Staff:[6]

>We believe that a considerable portion of the Japanese population now consider absolute military defeat to be probable. The increasing effects of sea blockade and cumulative devastation wrought by strategic bombing, which has already rendered millions homeless and has destroyed from 25 to 50 percent of the builtup area of Japan's most important cities, should make this realization increasingly general. An entry of the Soviet Union into the war would finally convince the Japanese of the inevitability of complete defeat. Although individual Japanese willingly sacrifice themselves in the service of the nation, we doubt that the nation as a whole is predisposed toward national suicide. Rather, the Japanese as a nation have a strong concept of national survival, regardless of the fate of individuals. They would probably prefer national survival, even through surrender, to virtual extinction.
>
>The Japanese believe, however, that unconditional surrender would be the equivalent of national extinction. There are as yet no indications that the Japanese are ready to accept such terms. The ideas of foreign occupation of the Japanese homeland, foreign custody of the person of the Emperor, and the loss of prestige entailed by the acceptance of 'unconditional surrender' are most revolting to the Japanese. *To avoid these conditions, if possible, and, in any event, to insure survival of the institution of the Emperor, the Japanese might well be willing to withdraw from all the territory they have seized on the Asiatic continent and in the southern Pacific, and even to agree to the independence of Korea and to the practical disarmament of their military forces.*
>
>A conditional surrender by the Japanese Government along the lines stated above might be offered by them at any time from now until the time of the complete destruction of all Japanese power of resistance.

7. Harry Truman had been President only three months when he attended the Potsdam Conference (July 17-August 2, 1945) as one of the "Big Three" in the formidable company of Winston Churchill and Joseph Stalin. On some important questions Truman tended to rely on advice tendered by Churchill, who was vastly more experienced in this arena. In the following excerpt from his book *Triumph and Tragedy*, the British Prime Minister discusses the advice he gave Truman on the problem of unconditional surrender, and states his confident expectations as to the intentions of American statesmen on this problem, expectations subsequently unfulfilled:[7]

>The professional diplomats [in Japan] were convinced that only immediate surrender under the authority of the Emperor could save Japan from complete disintegration, but power still lay almost entirely in the hands of

---

[6] *Ibid.*, II, pp. 36n.-37n.
[7] Winston S. Churchill, *Triumph and Tragedy* (Houghton Mifflin Co. and Cassell & Co., Ltd., Copyright 1953), pp. 641-642.

a military clique determined to commit the nation to mass suicide rather than accept defeat. . . .

In several lengthy talks with the President alone, or with his advisers present, I discussed what to do. Earlier in the week Stalin had told me privately that as his party was leaving Moscow an unaddressed message had been delivered to him through the Japanese Ambassador. . . . It stated that Japan could not accept "unconditional surrender," but might be prepared to compromise on other terms. Stalin had replied that as the message contained no definite proposals the Soviet Government could take no action. I explained to the President that Stalin had not wished to tell him direct lest he might think the Russians were trying to influence him towards peace. In the same way I thought we should abstain from saying anything which would make us seem at all reluctant to go on with the war against Japan for as long as the United States thought fit. However, I dwelt upon the tremendous cost in American and to a smaller extent in British life if we enforced "unconditional surrender" upon the Japanese. It was for him to consider whether this might not be expressed in some other way, so that we got all the essentials for future peace and security and yet left them some show of saving their military honour and some assurance of their national existence, after they had complied with all safeguards necessary for the conqueror. The President replied bluntly that he did not think the Japanese had any military honour after Pearl Harbour. I contented myself with saying that at any rate they had something for which they were ready to face certain death in very large numbers, and this might not be so important to us as it was to them. He then became quite sympathetic, and spoke, as had Mr. Stimson, of the terrible responsibilities that rested upon him for the unlimited effusion of American blood.

I felt there would be no rigid insistence upon "unconditional surrender," apart from what was necessary for world peace and future security and for the punishment of a guilty and treacherous deed. Mr. Stimson, General Marshall, and the President were evidently searching their hearts, and we had no need to press them. We knew of course that the Japanese were ready to give up all conquests made in the war.

8. The man who may have played the most crucial behind-the-scenes role in these discussions was Cordell Hull. He had been Secretary of State for eleven of the twelve years of Franklin D. Roosevelt's presidency, retiring in November, 1944, for reasons of ill health. In his *Memoirs,* **Hull stated:**[8]

Just before Secretary of State Byrnes left for the Potsdam Conference in July, 1945, he telephoned me at my apartment and gave me the substance of a draft statement which he said President Truman had given him. This proposed statement, for issue by the United States, Britain, and Russia at the Potsdam Conference, contained a declaration by the Allies to Japan that the Emperor institution would be preserved if Japan would make peace. Byrnes asked my opinion. He said that high officials of the State, War, and Navy Departments had approved it.

I replied that, since he was leaving in a few minutes, there was no time to write anything for him, but that the statement seemed too much like appeasement of Japan, especially after the resolute stand we had maintained

---

[8] Cordell Hull, *Memoirs* (New York, Macmillan, Copyright 1948), II, pp. 1593-1594.

on unconditional surrender. I pointed out that, as it was worded, it seemed to guarantee continuance not only of the Emperor but also of the feudal privileges of a ruling caste under the Emperor. I said that the Emperor and the ruling class must be stripped of all extraordinary privileges and placed on a level before the law with everybody else.

I then sent Byrnes a cable on July 16, through the courtesy of Under Secretary Grew, to outline my thoughts in further detail. I said that the support of the statement by the chief people in the State, War, and Navy Departments called for the most serious consideration. Nevertheless I pointed out that the central point calculated to create serious difference was in the paragraph relating to a proposed declaration by the Allies now—I underlined "now"—that the Emperor and his monarchy would be preserved in the event of an Allied victory. The proponents of this promise, I added, believed that somehow the influences and persons who paid allegiance to the Emperor and his religious status would fight and resist less hard and so save Allied lives and shorten the war.

The other side, however, I concluded, was that no person knew how the proposal would work out. The militarists would try hard to interfere. Also, should it fail, the Japanese would be encouraged and terrible political repercussions would follow in the United States. I therefore asked whether it would be well first to await the climax of Allied bombing and Russia's entry into the war.

The following day I received a message from Secretary Byrnes agreeing that the statement should be delayed, and that, when it was issued, it should not contain this commitment with regard to the Emperor.

9. One of President Truman's closest advisors, James F. Byrnes, was named Secretary of State early in July, 1945, on the eve of the Potsdam Conference. Fifteen years later in an interview published in *U. S. News and World Report,* **Byrnes defended the position he had taken on the question of unconditional surrender:**[9]

> *Q.* In retrospect, might it have been possible to avoid using the atom bomb by offering Japan a chance to keep its Emperor, as Joseph Grew . . . has stated?
>
> *A.* That's dealing in the realm of speculation. Later, on August 11, in drafting the message to Japan replying to their surrender message, I wrote that Japan would have the right to determine the form of government under which its people wished to live. It was approved by the President and by Stimson.
>
> *Q.* Would any assurance regarding the retaining of the institution of the Emperor have encouraged Japan to open negotiations for surrender sooner?
>
> *A.* I do not think so. The militarists were still in control. The record shows that on July 21, five days before the Potsdam Declaration was released, the Japanese Government advised its Ambassador in Moscow that "so long as the enemy demands unconditional surrender we will fight as one man against the enemy."
>
> *Q.* But in our final acceptance of their offer of surrender, didn't we agree to retain the institution of the Emperor? Wasn't that a change from the Potsdam Declaration?

---

[9] *U. S. News and World Report,* August 15, 1960, pp. 66-67.

A. No. When the Japanese Government submitted its agreement to surrender, provided the surrender did not envisage the insistence upon the removal of the Emperor, we replied that "from the moment of surrender the authority of the Emperor and the Japanese Government to rule the state shall be subject to the Supreme Commander of the Allied Powers, who will take such steps as he deems proper to effectuate the surrender terms."

Q. Did this represent any change of view on our part?

A. No, it did not. It was a requirement that the Emperor, as head of the Japanese Government, should agree to the terms of surrender. Then we added that it was for the people of Japan to determine the form of government under which they would live.

10. The final product of these many-sided deliberations was the Potsdam Declaration, issued on July 26, 1945. The words used in it—and even more significantly the words not used—have become one of the most controversial issues in modern history.[10]

> PROCLAMATION CALLING FOR THE SURRENDER OF JAPAN, APPROVED BY THE HEADS OF GOVERNMENT OF THE UNITED STATES, CHINA, AND THE UNITED KINGDOM...
>
> (1) We, the President of the United States, the President of the National Government of the Republic of China and the Prime Minister of Great Britain, representing the hundreds of millions of our countrymen, have conferred and agree that Japan shall be given an opportunity to end this war.
>
> (2) The prodigious land, sea and air forces of the United States, the British Empire and of China, many times reinforced by their armies and air fleets from the west are poised to strike the final blows upon Japan. This military power is sustained and inspired by the determination of all the Allied nations to prosecute the war against Japan until she ceases to resist.
>
> (3) The result of the futile and senseless German resistance to the might of the aroused free peoples of the world stands forth in awful clarity as an example to the people of Japan. The might that now converges on Japan is immeasurably greater than that which, when applied to the resisting Nazis, necessarily laid waste to the lands, the industry and the method of life of the whole German people. The full application of our military power, backed by our resolve, *will* mean the inevitable and complete destruction of the Japanese armed forces and just as inevitably the utter devastation of the Japanese homeland.
>
> (4) The time has come for Japan to decide whether she will continue to be controlled by those self-willed militaristic advisers whose unintelligent calculations have brought the Empire of Japan to the threshold of annihilation, or whether she will follow the path of reason.
>
> (5) Following are our terms. We will not deviate from them. There are no alternatives. We shall brook no delay.
>
> (6) There must be eliminated for all time the authority and influence of those who have deceived and misled the people of Japan into embarking on

---

[10] *Potsdam Papers.* II, pp. 1474-1476.

world conquest, for we insist that a new order of peace, security and justice will be impossible until irresponsible militarism is driven from the world.

(7) Until such a new order is established *and* until there is convincing proof that Japan's war-making power is destroyed, points in Japanese territory to be designated by the Allies shall be occupied to secure the achievement of the basic objectives we are here setting forth.

(8) The terms of the Cairo Declaration[11] shall be carried out and Japanese sovereignty shall be limited to the islands of Honshu, Hokkaido, Kyushu, Shikoku, and such minor islands as we determine.

(9) The Japanese military forces, after being completely disarmed, shall be permitted to return to their homes with the opportunity to lead peaceful and productive lives.

(10) We do not intend that the Japanese shall be enslaved as a race or destroyed as [a] nation, but stern justice shall be meted out to all war criminals, including those who have visited cruelties upon our prisoners. The Japanese government shall remove all obstacles to the revival and strengthening of democratic tendencies among the Japanese people. Freedom of speech, of religion, and of thought, as well as respect for the fundamental human rights shall be established.

(11) Japan shall be permitted to maintain such industries as will sustain her economy and permit the exaction of just reparations in kind, but not those industries which would enable her to re-arm for war. To this end, access to, as distinguished from control of raw materials shall be permitted. Eventual Japanese participation in world trade relations shall be permitted.

(12) The occupying forces of the Allies shall be withdrawn from Japan as soon as these objectives have been accomplished and there has been established in accordance with the freely expressed will of the Japanese people a peacefully inclined and responsible government.

(13) We call upon the Government of Japan to proclaim now the unconditional surrender of all the Japanese armed forces, and to provide proper and adequate assurances of their good faith in such action. The alternative for Japan is prompt and utter destruction.

11. One of Japan's most experienced diplomats, Shigenori Togo, served as Foreign Minister early in the war and again at its close. Togo's memoirs, published in 1956, furnish the following account of the frantic intrigues and debates that erupted within the Japanese government when the Potsdam Declaration was issued.[12]

My first reaction to the declaration upon reading through the text as broadcast by the American radio was that, in view of the language, "Following are our terms," it was evidently not a dictate of unconditional surrender. I got the impression that the Emperor's wishes had reached the United States and Great Britain, and had had the result of this moderation of their attitude. . . .

---

[11]Promulgated December 1, 1943 at a conference in Cairo, Egypt, by Roosevelt, Churchill and Generalissimo Chiang Kai-shek of China, this declaration was the first to demand Japan's unconditional surrender. It also pledged that Japan would be stripped of all her conquests since 1914.

[12]Shigenori Togo, *The Cause of Japan* (New York, Simon and Schuster), pp. 311-321, (c) 1956 by Simon and Schuster, Inc. Reprinted by permission.

Simultaneously, I thought it desirable to enter into negotiation with the Allied Powers to obtain some clarification, and revision—even if it should be slight—of disadvantageous points in the declaration.

I was received in audience on the morning of the 27th, and reported to the Emperor on recent happenings, including the negotiations with Moscow, the British general election[13] and the Potsdam Declaration. I stressed that the declaration must be treated with the utmost circumspection, both domestically and internationally; in particular, I feared the consequences if Japan should manifest an intention to reject it. I pointed out further that the efforts to obtain Soviet mediation to bring about the ending of the war had not yet borne fruit, and that our attitude toward the declaration should be decided in accordance with their outcome.

At a meeting of the members of the Supreme Council for Direction of the War, held on the same day, I spoke to the same effect. On this occasion, Chief of Staff Toyoda said that news of the declaration would, sooner or later, transpire, and if we did nothing it would lead to a serious impairment of morale; hence, he suggested, it would be best at this time to issue a statement that the government regarded the declaration as absurd and could not consider it. Premier Suzuki and I objected to this, and as a result it was agreed that for the time being we should wait to see what the response of the U.S.S.R. would be to our approach to her, planning to decide our course thereafter. On the same afternoon there was a Cabinet meeting, at which ... I went into detail concerning the Potsdam Declaration, and recommended that we should act on it after having ascertained the attitude of the Soviet Union. No dissent from this treatment of the declaration was expressed, though there was considerable discussion of the way and the extent of making it public. In the end it was agreed that it should be passed without comment by the government, the competent authorities releasing it in summary, while the Board of Information should lead the press to minimize publicity.

To my amazement, the newspapers of the following morning reported that the government had decided to ignore the Potsdam Declaration. I protested without delay to the Cabinet when it met, pointing out that the report was at variance with our decision of the preceding day. What had happened, I learned, was this. There had been held in the Imperial Palace, after adjournment of the Cabinet the day before, a conference for exchange of information between government and high command. This was a routine weekly meeting without special significance, and I had been absent because of more important business. One of the military participants in that meeting, as I heard it, had proposed the rejection of the Potsdam Declaration; the Premier, the War and Navy Ministers and the two Chiefs of Staff had hastily assembled for consultation in a separate room, and the Premier had been persuaded by the more militant elements to that course. He then stated at a subsequent press conference that the government had decided to ignore the declaration, and this announcement it was which the press had played up so sensationally. It was only after the affair had developed to this point that I first knew of it; despite my thorough dissatisfaction with the position, there was of course no way of withdrawing the statement released by the

---

[13] In the election held in Britain while the Potsdam Conference was in progress, Prime Minister Churchill and his Conservative Party were defeated. Thus after July 28 the leader of the Labour Party, Clement Attlee, replaced Churchill at Potsdam.

Premier, and things had to be left as they stood. In the result, the American press reported that Japan had rejected the declaration, and President Truman in deciding for use of the atomic bomb, and the U.S.S.R. in attacking Japan, referred to the rejection of it as justification for their respective actions. The incident was thus a deplorable one in its embarrassment of our move for peace, and was most disadvantageous for Japan....

At 8:15 A.M. on 6 August the United States Air Force released over Hiroshima the atomic bomb the detonation of which was to reverberate down through the history of the world. I was informed that the damage was vast. I immediately demanded of the Army the particulars; the American radio had announced that the bomb was one employing atomic fission, and if such a singular explosive had in fact been used, in violation of the international law of warfare, it would be necessary to lodge a protest with the United States. The Army replied to my inquiry that it could as yet say only that the bomb dropped on Hiroshima was one of high effectiveness, and that the details were under investigation. The United States and Great Britain launched large-scale propaganda on the atomic bomb, declaring that its use would alter utterly the character of war and would work a revolution in the life of the human race, and that if Japan did not accept the declaration of the three Powers the bomb would continue to be used until the nation was annihilated.

At a meeting of the Cabinet on the afternoon of 7 August the War and Home Mininsters made reports on the Hiroshima bombing. The Army, pleading the necessity of awaiting the results of the investigation which had been ordered, obviously intended not to admit the nature of the atomic attack, but to minimize the effect of the bombing. On the 8th I had an audience, in the underground shelter of the Imperial Palace, with the Emperor, whom I informed of the enemy's announcement of the use of an atomic bomb, and related matters, and I said that it was now all the more imperative that we end the war, which we could seize this opportunity to do. The Emperor approved of my view, and warned that since we could no longer continue the struggle, now that a weapon of this devastating power was used against us, we should not let slip the opportunity by engaging in attempts to gain more favorable conditions. Since bargaining for terms had little prospect of success at this stage, he said, measures should be concerted to insure a prompt ending of hostilities. He further added that I should communicate his wishes to the Premier....

In the early hours of the 9th the radio room of the Foreign Ministry telephoned to inform me of the U.S.S.R.'s broadcast of her declaration of war on us and the large-scale invasion of Manchuria by her forces....

The members of the Supreme Council met at 11:00 A.M. I opened the discussion by saying that the war had become more and more hopeless, and now that it had no future, it was necessary to make peace without the slightest delay. Therefore, I said, the Potsdam Declaration must be complied with, and the conditions for its acceptance should be limited to those only which were absolutely essential for Japan. All members of the Supreme Council already recognized the difficulties in going on with the war; and now, after the employment of the atomic bomb and Russian entry into the war against us, none opposed in principle our acceptance of the declaration. None disagreed, either, that we must insist upon preservation of the national polity [the position of the Emperor] as the indispensable condition of acceptance.

The military representatives, however, held out for proposing additional terms—specifically, that occupation of Japan should if possible be avoided or, if inescapable, should be on a small scale and should not include such points as Tokyo; that disarmament should be carried out on our responsibility; and that war criminals should be dealt with by Japan. I objected that in view of the recent attitude of Britain, America, Russia and China it was greatly to be feared that any proposal by us of a number of terms would be rejected, and that the entire effort for peace would be in danger of failing. Unless, therefore, the military services saw a prospect of winning the war, any terms proposed by us should be limited to the minumum of those truly vital; thus while it was in order to propose other points as our desire, the only condition as such which we should hold out for was that of inviolability of the Imperial house. I asked, then, whether the armed services could offer any hope of victory in case negotiations on terms should be undertaken and should fail.

The War Minister replied that although he could give no assurance of ultimate victory, Japan could still fight another battle. I pressed them to say whether they could be certain of preventing the enemy from landing on our mainland. The Army Chief of Staff answered that we might drive the enemy into the sea if all went very well—though, in war, we could not be confident that things would go well—but that even conceding that a certain percentage of the enemy's troops might succeed in establishing beachheads, he was confident that we could inflict heavy losses on them. I argued that this would be of no avail; according to the explanation given us by the Army, some part at least of the attackers might still land even after sustaining serious losses; but while it was obvious that the enemy would follow up with a second assault though the first was inadequately rewarded, we should have sacrificed most of our remaining aircraft and other important munitions in our efforts to destroy the first wave. There being no possibility of replenishing our supply of armaments in a short period, our position after the first enemy landing operations would be one of defenselessness, even leaving the atomic bomb out of account. My conclusion was that we had no alternative to stopping the war at this very moment, and we must therefore attempt to attain peace by limiting our counterdemands to the irreducible minimum.

The discussion became rather impassioned, but remained inconclusive, and it neared one o'clock, with a Cabinet meeting scheduled for the afternoon. The Premier stated that the question had to be submitted to the Cabinet also, and the Supreme Council adjourned without having come to any agreement how we should proceed.

The Cabinet met at two.... I again detailed the course of the negotiations with the U.S.S.R., the use of the atomic bomb and the Soviet attack on us. There was the same controversy.... The meeting had gone for hours, and it was now late at night. The Premier asked the Cabinet members to state their conclusions; some equivocated, some agreed with the Army's view, but most supported me.

At that point the Premier stated that he wished to report to the Emperor with me alone. Leaving the Cabinet in session, we went together to the Palace. Upon our being received, the Premier requested that I outline to the Emperor the disagreement in the Supreme Council and the Cabinet, which I did fully. The Premier than asked the Emperor's sanction for calling at once, that night, a meeting in his presence of the Supreme Council for

Direction of the War. The Emperor approved, and the Imperial Conference convened shortly before midnight of the 9th.

The Premier opened the conference by saying that, the deliberations at that morning's Supreme Council having failed to result in agreement on the accepting of the Potsdam Declaration, he wished to ask the Emperor to hear personally the opposing views. Thereupon two alternatives were submitted for consideration: one, to accept the Potsdam Declaration with the understanding that it comprised no demand which would prejudice the traditionally established status of the Emperor; the other, to attach in addition the three conditions before mentioned as insisted upon by the Army. I dilated upon the same points which I had argued at the Supreme Council members' meeting and that of the Cabinet, and contended that we must now end the war by accepting the Potsdam Declaration in accordance with the first alternative. The Navy Minister said simply that he fully concurred in the opinion of the Foreign Minister. But War Minister Anami reiterated his argument that we should propose the additional conditions, and the Army Chief of Staff announced a similar conviction. Baron Hiranuma [President of the Privy Council], after having asked a number of questions, called for amendment of the reservation in the first alternative to provide that the declaration "comprised no demand which would prejudice the prerogatives of the Emperor as a sovereign ruler"; this amendment being approved by all, Hiranuma agreed to that alternative.

There being still a division of opinion, the Premier said that he regretted that he must humbly beg the Emperor's decision. The Emperor quietly said that he approved the opinion of the Foreign Minister; the confidence of the services in ultimate victory, he said, could not be relied upon, their earlier forecasts having often been at variance with the realities.... Now, bearing the unbearable, he would submit to the terms of the Potsdam Declaration, thereby to preserve the national polity.

The Imperial Conference thereupon ended, at about half-past two. The Cabinet met at 3:00 A.M., and unanimously adopted a decision in conformity with the Emperor's words.

B. THE RUSSIAN PROBLEM

From the Bolshevik Revolution of 1917, which established communist power in Russia, to 1945, Soviet-American relations were seldom cordial. In 1919 the United States joined the other great powers in sending troops to support the anti-Bolshevik forces in Russia, an act still bitterly resented by the communists. The United States was also one of the last major nations to recognize the Soviet government, not doing so until 1933. After the German invasion of Russia in June 1941, however, and especially after American entry into the war six months later, military strategy dictated the provision of massive aid to our new Russian ally in the common struggle against Nazi Germany. American policy was also aimed at least as early as 1943 at obtaining a Soviet pledge to enter the war against Japan as soon as Germany was beaten. But once Germany had surrendered and the western Allies had begun to work closely with Soviet Russia in the military occupation of central Europe, new considerations came to the fore.

Since the war a number of writers have asserted that America's true motives for using the atomic bomb were more anti-Soviet than anti-Japanese. From time to time this charge of "atomic blackmail" is repeated by the Russians and has been echoed by others throughout the world.

1. In 1948, Patrick M. S. Blackett, one of Britain's foremost atomic scientists, won the Nobel Prize in physics. He has since written and lectured widely on nuclear weapons control. The book from which this selection is taken was written in 1948:[14]

> Since the next major United States move was not to be until November 1, clearly there was nothing in the Allied plan of campaign to make urgent the dropping of the first bomb on August 6 rather than at any time in the next two months. Mr. Stimson himself makes clear that had the bombs not been dropped, the intervening period of eleven weeks between August 6 and the invasion planned for November 1 would have been used to make further fire raids with B–29's on Japan. Under conditions of Japanese air defense at that time, these raids would certainly have led to very small losses of American air personnel.
>
> Mr. Stimson's hurry becomes still more peculiar since the Japanese had already initiated peace negotiations....
>
> A plausible solution of this puzzle of the overwhelming reasons for urgency in the dropping of the bomb is not, however, hard to find....
>
> The U.S.S.R. declared war on Japan on August 8, and their offensive started early on August 9. On August 24, the Soviet High Command announced that the whole of Manchuria, Southern Sakhalin, etc., had been captured and that the Japanese Manchurian army had surrendered. No doubt the capitulation of the home government on August 14 reduced the fighting spirit of the Japanese forces. If it had not taken place, the Soviet campaign might well have been more expensive; but it would have been equally decisive. If the saving of American lives had been the main objective, surely the bombs would have been held back until *(a)* it was certain that the Japanese peace proposals made through Russia were not acceptable, and *(b)* the Russian offensive, which had for months been part of the Allied strategic plan, and which Americans had previously demanded, had run its course....
>
> As far as our analysis has taken us we have found no compelling military reason for the clearly very hurried decision to drop the first atomic bomb on August 6, rather than on any day in the next six weeks or so. But a most compelling diplomatic reason, relating to the balance of power in the postwar world, is clearly discernible.
>
> Let us consider the situation as it must have appeared in Washington at the end of July, 1945. After a brilliant, but bitterly-fought campaign, American forces were in occupation of a large number of Japanese islands. They had destroyed the Japanese Navy and Merchant Marine and largely destroyed their Air Force and many divisions of their Army; but they had still not come to grips with a large part of the Japanese land forces. Supposing the bomb had not been dropped, the planned Soviet offensive in Manchuria, so long demanded and, when it took place, so gladly welcomed (officially), would have achieved its objective according to plan. This must have been clearly foreseen by the Allied High Command, who knew well the great superiority of the Soviet forces in armor, artillery and aircraft, and who could draw on the experience of the European war to gauge the probable success of such a well-prepared offensive. If the bombs had not been dropped, America would have seen the Soviet armies engaging a major part

---

[14]Patrick M. S. Blackett, *Fear, War and the Bomb* (New York, McGraw-Hill, 1949), pp. 130-131, 132, 134-135, 139.

of Japanese land forces in battle, overrunning Manchuria and taking half a million prisoners. And all this would have occurred while American land forces would have been no nearer Japan than Iwo Jima and Okinawa. One can sympathize with the chagrin with which such an outcome would have been regarded. Most poignantly, informed military opinion could in no way blame Russia for these expected events. Russia's policy of not entering the war till Germany was defeated was not only military common sense but part of the agreed Allied plan.

In this dilemma, the successful explosion of the first atomic bomb in New Mexico, on July 16, must have come as a welcome aid. One can imagine the hurry with which the two bombs—the only two existing—were whisked across the Pacific to be dropped on Hiroshima and Nagasaki just in time, but only just, to insure that the Japanese government surrendered to American forces alone. The long-demanded Soviet offensive took its planned victorious course, almost unheralded in the world sensation caused by the dropping of the bombs....

So we may conclude that the dropping of the atomic bombs was not so much the last military act of the second World War, as the first major operation of the cold diplomatic war with Russia now in progress....

2. The author of this 1965 version of the anti-Soviet charge has worked in Washington, D. C., for a number of years as a Congressional legislative assistant and as Special Assistant in the State Department:[15]

Although military considerations were not primary, as we have seen, unquestionably political considerations related to Russia played a major role in the decision; from at least mid-May American policy makers hoped to end the hostilities before the Red Army entered Manchuria. For this reason they had no wish to test whether Russian entry into the war would force capitulation—as most thought likely—long before the scheduled November invasion. Indeed, they actively attempted to delay Stalin's declaration of war.

Nevertheless, it would be wrong to conclude that the atomic bomb was used simply to keep the Red Army out of Manchuria....

Such a conclusion is very difficult to accept, for American interests in Manchuria . . . were not of great significance. The further question therefore arises: Were there other political reasons for using the atomic bomb? In approaching this question, it is important to note that most of the men involved at the time who since have made their views public always mention *two* considerations which dominated discussions. The first was the desire to end the Japanese war quickly, which, as we have seen, was not primarily a military consideration, but a political one. The second is always referred to indirectly....

In essence, the second of the two overriding considerations seems to have been that a combat demonstration was needed to convince the Russians to accept the American plan for a stable peace....

Why did the American government refuse to attempt to exploit Japanese efforts to surrender? Or, alternatively, why did they refuse to test whether

---

[15] Gar Alperovitz, *Atomic Diplomacy: Hiroshima and Potsdam* (New York, Simon & Schuster, Copyright (c) by Gar Alperovitz, 1965), pp. 239-242. Reprinted by permission of Simon & Schuster and Martin Secker & Warburg Ltd.

a Russian declaration of war would force capitulation? Were Hiroshima and Nagasaki bombed primarily to impress the world with the need to accept America's plan for a stable and lasting peace—that is, primarily, America's plan for Europe? The evidence strongly suggests that the view which the President's personal representative offered to one of the atomic scientists in May 1945 was an accurate statement of policy: "Mr. Byrnes did not argue that it was necessary to use the bomb against the cities of Japan in order to win the war.... Mr. Byrnes's ... view [was] that our possessing and demonstrating the bomb would make Russia more manageable in Europe...."

3. An American historian who has written a series of detailed studies of Far Eastern diplomatic relations in World War II here attempts to strike a balance between the charge and the evidence:[16]

Some of those men who concurred in the decision to use the bomb discerned other advantages and justifications. It may be also—but this is only conjecture—that Churchill and Truman and some of their colleagues conceived that besides bringing the war to a quick end, it would improve the chances of arranging a satisfactory peace. For would not the same dramatic proof of western power that shocked Japan into surrender impress the Russians also? Might it not influence them to be more restrained? Might it not make more effective the resistance of the western allies to excessive Soviet pretensions and ventures, such as the Soviet bids for a military base in the Black Sea Straits, and a part of the occupation and control of Japan akin to that which it had in Germany? If these conjectures have any basis in actuality, they would provide another justification for using the bomb as a military weapon. We were not only subduing Japanese aggressors; we were perhaps monitoring the emergent Russian aggression.

Recognition of this tendency must not be distorted into an accusation that the American government engaged in what Soviet propagandists and historians have called "atomic blackmail." To the contrary, the American government remained intently desirous of preserving the friendly connection with the Soviet Union. It had rejected Churchill's proposal to face down the Soviet government in some climactic confrontation over the outward thrust of Soviet power. It showed throughout the Conference at Potsdam, which preceded the use of the bomb, a patient disposition to arrive at compromises with the Soviet government. In brief, the purposes of the men who determined American policy were directed toward achieving a stable international order by peaceful ways, not swayed by excited thought or wish of imposing our will on the rest of the world by keeping atomic bombs poised over their lives.

4. On the question of Soviet-American relations, as on so many others, Secretary Stimson was among the first to consider post-war as well as immediate implications. His concern about this subject is revealed in his memoirs:[17]

---

[16]Herbert Feis, *The Atomic Bomb and the End of the War in the Pacific* originally published as *Japan Subdued* (Princeton, Princeton University Press, Copyright 1961), pp. 181-182.
[17]Henry L. Stimson and McGeorge Bundy, *On Active Service In Peace and War,* pp, 637-641.

Even the immediate tactical discussion about the bomb involved the Russians. Much of the policy of the United States toward Russia, from Teheran to Potsdam, was dominated by the eagerness of the Americans to secure a firm Russian commitment to enter the Pacific war. And at Potsdam there were Americans who thought still in terms of securing Russian help in the Pacific war. Stimson himself had always hoped that the Russians would come into the Japanese war. . . .

The news from Alamogordo, arriving at Potsdam on July 16, made it clear to the Americans that further diplomatic efforts to bring the Russians into the Pacific war were largely pointless. The bomb as a merely probable weapon had seemed a weak reed on which to rely, but the bomb as a colossal reality was very different. The Russians may well have been disturbed to find that President Truman was rather losing his interest in knowing the exact date on which they would come into the war.

The Russians at Potsdam were not acting in a manner calculated to increase the confidence of the Americans or the British in their future intentions. . . . Naturally, therefore, news of the atomic bomb was received in Potsdam with great and unconcealed satisfaction by Anglo-American leaders. At first blush it appeared to give democratic diplomacy a badly needed "equalizer."

Stimson personally was deeply disturbed, at Potsdam, by his first direct observation of the Russian police state in action. The courtesy and hospitality of the Russians was unfailing, but there was evident nonetheless, palpable and omnipresent, the atmosphere of dictatorial repression. Nothing in his previous life matched this experience, and it was not particularly heartening to know that the Soviet machine for the time being was operating to insure the comfort and safety of the Allied visitors. Partly at firsthand and partly through the reports of Army officers who had observed the Russians closely during the first months of the occupation, Stimson now saw clearly the massive brutality of the Soviet system and the total suppression of freedom inflicted by the Russian leaders first on their own people and then on those whose lands they occupied. The words "police state" acquired for him a direct and terrible meaning. What manner of men were these with whom to build a peace in the atomic age?

For the problem of lasting peace remained the central question. Any "equalizing" value of the atomic bomb could only be of short-range and limited value, however natural it might be for democratic leaders to be cheered and heartened by the knowledge of their present possession of this final arbiter of force. As Stimson well knew, this advantage was temporary.

But could atomic energy be controlled, he asked himself, if one of the partners in control was a state dictatorially and repressively governed by a single inscrutable character? Could there be *any* settlement of lasting value with the Soviet Russia of Stalin? With these questions and others crowding his mind, he wrote in Potsdam for the President a paper headed, "Reflections on the Basic Problems Which Confront Us." . . .

The central concern of this paper was the Russian police state, and only secondly the atomic bomb. Stimson's first main point was that the present state of Russia, if continued without change, would very possibly in the end produce a war. . . .

It was easier to state the problems and insist that it be solved than to suggest any course likely to be effective. . . . And in the last paragraph of his Potsdam relections Stimson came to a gloomy conclusion.

"7. The foregoing has a vital bearing upon the control of the vast and revolutionary discovery of X [atomic energy] which is now confronting us. Upon the successful control of that energy depends the future successful development or destruction of the modern civilized world. The committee appointed by the War Department which has been considering that control has pointed this out in no uncertain terms and has called for an international organization for that purpose. After careful reflection I am of the belief that *no* world organization containing as one of its dominant members a nation whose people are not possessed of free speech, but whose governmental action is controlled by the autocratic machinery of a secret political police, can give effective control of this new agency with its devastating possibilities.

"I therefore believe that . . . we must go slowly in any disclosures or agreeing to any Russian participation whatsoever and constantly explore the question how our headstart in X and the Russian desire to participate can be used to bring us nearer to the removal of the basic difficulties which I have emphasized."

5. Shortly before James Byrnes was appointed Secretary of State, early in July, on the eve of the Potsdam Conference, he had a meeting with three atomic scientists. The group was headed by Leo Szilard, the physicist who had helped launch the Manhattan Project and who was now a leader in the scientists' movement to deter the government from using the bomb. According to Szilard's account, from which the following is excerpted, the interview provided significant testimony relating to the Russian problem:[18]

The question of whether the bomb should be used in the war against Japan came up for discussion. Mr. Byrnes did not argue that it was necessary to use the bomb against the cities of Japan in order to win the war. He knew at that time, as the rest of the Government knew, that Japan was essentially defeated and that we could win the war in another six months. At that time Mr. Byrnes was much concerned about the spreading of Russian influence in Europe. . . . Mr. Byrnes's concern about Russia I fully shared, but his view that our possessing and demonstrating the bomb would make Russia more manageable in Europe I was not able to share. Indeed I could hardly imagine any premise more false and disastrous upon which to base our policy, and I was dismayed when a few weeks later I learned that he was to be our Secretary of State.

6. Byrnes never made any secret of his feelings about the Russians, as his memoirs amply demonstrate. The two incidents he describes in these excerpts took place during the Potsdam Conference:[19]

Secretary Forrestal arrived and told me in detail of the intercepted messages from the Japanese government to Ambassador Sato in Moscow, indicating Japan's willingness to surrender. He also quoted General Eisenhower as saying that President Truman had told him his principal objective at Potsdam would be to get Russia in the war. I told him it was most

---

[18]Leo Szilard, "A Personal History of the Bomb," *The Atlantic Community Faces the Bomb*, University of Chicago Round Table No. 601, September 25, 1949, pp. 14-15.
[19]James F. Byrnes, *All in One Lifetime* (Harper & Row, Copyright (c) 1958 by James F. Byrnes Foundation), pp. 297, 300-301.

probable that the President's views had changed; certainly that was not now my view. Forrestal replied that he thought it would take an army to keep Stalin out, with which I agreed....

The President and I discussed whether or not we were obligated to inform Stalin that we had succeeded in developing a powerful weapon and shortly would drop a bomb in Japan. Though there was an understanding that the Soviets would enter the war with Japan three months after Germany surrendered, which would make their entrance about the middle of August, with knowledge of the Japanese peace feeler and the successful bomb test in New Mexico, the President and I hoped that Japan would surrender before then. However, at luncheon we agreed that because it was uncertain, and because the Soviets might soon be our allies in that war, the President should inform Stalin of our intention, but do so in a casual way.

He then informed the British of our plan, in which they concurred. Upon the adjournment of the afternoon session, when we arose from the table, the President, accompanied by our interpreter, Bohlen, walked around to Stalin's chair and said, substantially, "You may be interested to know that we have developed a new and powerful weapon and within a few days intend to use it against Japan." I watched Stalin's expression as this was being interpreted, and was surprised that he smiled blandly and said only a few words. When the President and I reached our car, he said that the Generalissimo had replied only, "That's fine. I hope you make good use of it against the Japanese."

I did not believe Stalin grasped the full import of the President's statement, and thought that on the next day there would be some inquiry about this "new and powerful weapon," but I was mistaken.

I thought then and even now believe that Stalin did not appreciate the importance of the information that had been given him; but there are others who believe that in the light of later information about the Soviets' intelligence service in this country, he was already aware of the New Mexico test, and that this accounted for his apparent indifference.

7. A previously cited excerpt from Churchill's memoirs expressed his delight when the news of the successful Alamagordo test of the atomic bomb was received at Potsdam. Greatly relieved as he was at this elimination of the necessity for the invasion of Japan, he had other grounds for satisfaction, as this excerpt reveals:[20]

Moreover, we should not need the Russians. The end of the Japanese war no longer depended upon the pouring in of their armies for the final and perhaps protracted slaughter. We had no need to ask favours of them. A few days later I minuted to Mr. Eden: "It is quite clear that the United States do not at the present time desire Russian participation in the war against Japan." The array of European problems could therefore be faced on their merits and according to the broad principles of the United Nations. We seemed suddenly to have become possessed of a merciful abridgment of the slaughter in the East and of a far happier prospect in Europe. I have no doubt that these thoughts were present in the minds of my American friends.

---

[20]Winston Churchill, *Triumph and Tragedy*, p. 639.

8. Harry Truman had been President eleven days when he faced his first crisis with the Soviet Union. Having learned that the Russians had repudiated certain agreements concluded at the Yalta Conference two months earlier, he called a conference of his key advisors on April 23, 1945. Charles Bohlen, Special Assistant to the Secretary of State, describes the President's vigorous reaction:[21]

> The President said . . . that he felt our agreements with the Soviet Union so far had been a one-way street and that he could not continue; it was now or never. He intended to go on with the plans for San Francisco[22] and if the Russians did not wish to join us they could go to hell.

9. In mid-July General Dwight D. Eisenhower, at that time Supreme Commander of the Allied forces in Europe, conferred informally with the President at Potsdam. He recounted part of their conversation in *Crusade in Europe*, a memoir of the war years:[23]

> Another item on which I ventured to advise President Truman involved the Soviets' intention to enter the Japanese war. I told him that since reports indicated the imminence of Japan's collapse I deprecated the Red Army's engaging in that war. I foresaw certain difficulties arising out of such participation and suggested that, at the very least, we ought not to put ourselves in the position of requesting or begging for Soviet aid. It was my personal opinion that no power on earth could keep the Red Army out of that war unless victory came before they could get in.

---

[21] Walter Millis, ed., *The Forrestal Diaries* (New York, Viking Press, 1951), p. 50.
[22] The Conference then being planned was expected to draw up the charter for the United Nations.
[23] Dwight D. Eisenhower, *Crusade in Europe* (New York, Doubleday 1948), pp. 441-442.

SECTION 5

Was It a Morally Defensible Act?

The American conscience has never ceased to be troubled by the ethical dilemma posed by Hiroshima. As we have seen, many of those who participated in the making of the bomb and in the decision to use it have felt obligated to justify their actions. In the ensuing years others have sought to analyze the problem in the light of generally accepted moral principles.

A. PUBLIC CONSCIENCE IN A DECADE OF WAR

On August 30, 1936, in the early stages of the Spanish Civil War, a single plane bombed Madrid. There were casualties, but no one was killed. The world was nonetheless horrified at the thought of a great metropolitan center being attacked from the air.

Nine years later, it was also a single plane that bombed Hiroshima.

Between these two events lies the bloodiest era in human history. The conscience of mankind was ceaselessly assailed by steadily mounting horrors in every quarter of the globe. Hiroshima cannot be properly understood outside this historical setting.

1. Following is a selection of page 1 headlines and editorial comment taken from *The New York Times* over the period **1936-1945**. All have to do with the bombing of cities. They reflect not only dramatically changing circumstances in the period, but changing American attitudes toward those circumstances.[1]

In 1936 a bloody civil war broke out in Spain, with the forces of Francisco Franco eventually toppling the republican government.

August 30, 1936:
>   REBEL FLIER BOMBS HEART OF MADRID;
>   17 ARE WOUNDED

October 31, 1936:
>   REBEL PLANES DROP BOMBS IN MADRID;
>   110 KILLED, 135 HURT

---

[1] All headlines are taken from page 1 of *The New York Times* unless otherwise indicated. The dates are generally the day following the events described.

November 1, 1936:
(editorial)

"NEUTRALITY" IN SPAIN

Non-intervention in Spain has become a cruel mockery. Behind the screen of neutrality scores of women and children were slaughtered last Friday in the streets and playgrounds of Madrid, reportedly by foreign bombers in the service of the Rebels—a grisly rebuttal to the angry protests of innocence on the part of the Fascist Powers. . . .

December 3, 1936:

NEW MADRID RAIDS . . .
BOMBS RAINED ON CAPITAL
Toll Is Put in Hundreds in the
Rebels' Fiercest Air Attack

In mid-1937, Japan launched a full-scale though undeclared war against China.

September 9, 1937:

JAPANESE BOMBERS KILL 300 REFUGEES
FLEEING ON A TRAIN
400 Others Wounded in Raid Near Shanghai
By 5 Planes—Women and Children Victims

September 20, 1937:

PLANES FIGHT AT NANKING IN
BIG RAIDS BY JAPANESE . . .
2 ATTACKS ON CITY
Invaders Claim a Heavy Toll . . .

September 23, 1937:

U. S. IN SHARP NOTE TO JAPAN
'OBJECTS' TO NANKING RAIDS;
FIFTY PLANES ATTACK CITY
CASUALTIES AT 200

September 24, 1937:

20 CHINESE CITIES BOMBED:
2,000 CASUALTIES IN CANTON

September 24, 1937:
(editorial)

BOMBS OVER CHINA

Hundreds of dead and dying Chinese, mangled by high explosives in the narrow streets of Nanking and Canton, mock the bland assurances which the Japanese Government has given. . . .

In the protest sent by our own Government to Tokyo . . . the dominating note was one of sheer incredulity. The American Government simply "could not believe" . . . that the proposal to subject the whole Nanking area to bombing operations really "represents the considered intent of the Japanese Government." . . . Aimed against "an extensive area wherein there resides a large populace engaged in peaceful pursuits," the bombing of Nanking would be "unwarranted and contrary to principles of law and of humanity." Yet it has happened. . . .

These bombings of defenseless cities have been undertaken, said a spokesman of the Japanese Navy at Shanghai yesterday, "in order to bring the war to an early conclusion . . ."

September 26, 1937:
> 80 JAPANESE PLANES BOMB NANKING FOR
> SEVEN HOURS; 200 KILLED; HEAVY DAMAGE

World War II is usually said to have begun with the Nazi invasion of Poland in 1939. It was here that the world first learned the meaning of the German term **blitzkreig,** *"lightning war."*

September 1, 1939:
> GERMAN ARMY ATTACKS POLAND;
> CITIES BOMBED

September 2, 1939:
(editorial)
> TRAGEDY IN EUROPE

The first feeling of a heartsick world will be sheer inability to believe that the thing long feared has actually happened. . . . Here is war . . . Here are cities like our own, cities filled with men and women who might be our friends, bombed from the air with high explosives. . . .

September 3, 1939:
> 21 CIVILIANS KILLED
> IN RAID ON WARSAW
> Women, Children Die As Bomb
> Hits Workers' Apartment

September 5, 1939:
> FIRES IN WARSAW
> Nazi Bombers Terrorize City in
> 2 Raids—Many Slain, Damage Heavy

By the summer of 1940 the Germans had conquered Poland and virtually all of western Europe. England stood alone. Now began the "Battle of Britain," in which Hitler attempted to destroy Britain's will to resist the invasion he planned for her.

August 16, 1940:
> 1,000 NAZI PLANES RAID BRITAIN

August 25, 1940:
> LONDON BOMBED IN 4 AREAS . . .
> First London Raid Starts Fires

August 28, 1940:
> NAZIS STRIKE AT 20 CITIES

August 29, 1940:
1,000 INCENDIARY BOMBS RAINED ON LONDON IN CITY'S LONGEST RAID

September 8, 1940:
1,500 NAZI PLANES BOMB LONDON

September 29, 1940:
NAZIS IN 4TH WEEK OF LONDON RAIDS . . . FLAMES IN CAPITAL

The Royal Air Force withstood these onslaughts of the German Luftwaffe, forcing cancellation of the invasion of Britain. It was, of course, bombing from the air at Pearl Harbor that brought the United States into the war.

Dec. 8, 1941:
JAPAN WARS ON U.S. AND BRITAIN . . .
Tokyo Bombers Strike Hard At
Our Main Bases on Oahu

Dec. 8, 1941:
(editorial)
WAR WITH JAPAN
We will reply with our full force, without panic and without losing sight of our objectives. . . . But in making war on Japan we will not overestimate the ability of Japan to do us harm; we will not mistake the lesser danger for the greater danger, and we will not forget that Hitler, and not Tokyo, is the greatest threat to our security. The real battle of our times will not be fought in the Far East. . . .

December 26, 1941:
MANILA DECLARED OPEN CITY

December 28, 1941:
JAPANESE BOMBS FIRE OPEN CITY OF MANILA; CIVILIAN TOLL HEAVY

December 28, 1941:
(editorial)
THE BOMBING OF MANILA
On Wednesday of last week Manila was made an "open city.". . .

Yet Manila was bombed. It was bombed despite the fact that . . . no defense was offered when the Japanese planes came over. It was bombed hour after hour; bombed mercilessly; bombed brutally; with machine-gun fire from the planes directed from low altitude at civilians as they fled for shelter.

That the same Japanese regime which stabbed us in the back at Pearl Harbor . . . should . . . add still another chapter to a record of infamy that includes the rape of Nanking, need not surprise us. . . .

Japan has made another miscalculation . . . She will have reason to remember the bombing of Manila.

American reaction to the Japanese attack had to be limited to a holding action for some time, as the major war effort was aimed at Germany. By mid-1942 the Germans were on the defensive in the air. First the Royal Air Force mounted a series of retaliatory raids, joined eventually by the U. S. Eighth Air Force based in Britain.

June 1, 1942:
>1,000 BRITISH BOMBERS SET COLOGNE ON FIRE

June 2, 1942:
>COLOGNE DEATH TOLL PUT AS HIGH AS 20,000
>WITH 54,000 HURT; NAZIS RAID CANTERBURY . . .
>REPRISAL DIRECTED AT CATHEDRAL CITY

June 2, 1942:
(editorial)
>THE TIDE OF WAR: 1940-42

Two years. Two years of horror, of tragedy, of sorrow. The battleline has widened. It is the whole earth. . . . It is democracy no longer on the defensive, grimly standing by while the Nazi planes come over London, Plymouth and Coventry, but getting control of the air over Western Europe, striking with a thousand planes at Cologne, preparing to send 2,000, to send 3,000, to send enough to paralyze the Nazi will and ability to fight. . . .

We can go on . . . only in the hope that these dreadful things may not have to be done again . . . that in this horrible bath of total war we shall purge the world of war and the makers of war. . . .

July 28, 1942:
>RAID LEAVES HAMBURG AFIRE
>600 R.A.F. Planes Strike Searing
>Blow at Reich Port
>175,000 Incendiaries and Many 2-Ton
>Bombs Loosed in 50-Minute Attack

May 18, 1943:
>R.A.F. BLASTS 2 BIG DAMS IN REICH;
>RUHR POWER CUT, TRAFFIC HALTED
>AS FLOODS CAUSE DEATH AND RUIN

July 31, 1943:
>BOMBS SMASH HAMBURG AND KASSEL AGAIN

August 2, 1943:
>BERLIN EVACUATION ORDERED;
>HAMBURG'S FATE STIRS FEARS

August 25, 1943:
>700-PLANE RAID OPENS BATTLE OF BERLIN

August 25, 1943:
(editorial)

BERLIN EXPRESS

What has already happened to Hamburg began to happen to Berlin just before midnight on Monday. . . .

The raid doesn't add up to anything that civilians safe at home in an unbombed country can greet with a light-hearted cheer. It is not pleasant to get at Hitler and his crew by hitting comparatively innocent civilians. . . .

Yet, with all its horrors and all its bitter cost, attack by air is less horrible and less costly than attack on the ground. . . . The air attack carries this war home to those who made it. . . .

April 30, 1944:

1,000 U. S. 'HEAVIES' BLAST BERLIN

In mid-1944 the world faced a new horror as the Germans became the first nation to use ballistic missiles against a city.

July 7, 1944:

LONDON IS FLYING BOMB TARGET, 2752 KILLED, CHURCHILL REVEALS . . .
Robots Hurt 8,000

The Allied air assault on Germany continued without letup until the Nazis surrendered on May 8, 1945. Meanwhile, American forces were advancing against the Japanese in the western Pacific. In the summer of 1944, from air bases in the newly captured Marianas, giant new B-29 Superfortresses launched the mightiest aerial onslaught of all time upon the Japanese homeland. It mounted towards its climax in the spring of 1945.

March 10, 1945:

300 B-29'S FIRE 15 SQUARE MILES OF TOKYO . . .
B-29's Pour Over 1,000 Tons of Incendiaries
on Japanese Capital . . .
Tremendous Fires Leap Up in Thickly Populated
Center of Big City

March 12, 1945:

B-29'S BLAST NAGOYA IN 2D
BIG BLOW IN 2 DAYS . . .
Low-Level Strike Equals That At Tokyo

March 12, 1945:
(editorial)

TOKYO IN FLAMES

The devastated district . . . was one of the most congested districts in the world. . . . A genuine conflagration, which is what hit Tokyo, leaves little above ground but the surface of the earth. . . .

The flames merged in one roaring mass. . . . Currents of hot air scorched our big planes at 5,000 feet and tossed them about like chips. The burning city served as a beacon to approaching planes 200 miles away. . . .

All this sounds as if Tokyo might not be an ash heap. ... We will certainly send our flying fleets again and again over Tokyo and Nagoya. These great cities may become no more than holes in the ground. But so far in this war the Japanese have done their deadliest fighting from holes in the ground.

March 14, 1945:
> NEW B-29 BLOW FIRES 4 SQUARE
> MILES OF OSAKA ...
> Damage Is Called Worse Than Nagoya's ...
> 300 Planes ... Strike Mainland's 2d City

March 15, 1945:
(editorial)
> TRIAL BY FIRE

Even more important than the actual damage done would be the information, if we could get it, as to how the Japanese civilians reacted to these mass fire attacks on their great cities. ... They are not the stoics they sometimes are pictured as being. In a crisis they often panic. ...

The cost of the attack on Tokyo was two planes lost. The other strikes probably saw no greater toll, as similar conditions to those at Tokyo were reported—inaccurate anti-aircraft fire ... and timid fighter plane interception. ...

The Japanese can look forward to ever more frequent and ever heavier attacks.

March 17, 1945:
> B-29'S SET 12 SQUARE MILES OF KOBE AFIRE

May 24, 1945:
> 550 B-29'S SET TOKYO FIRES ...
> Shinagawa Section Is Target of 700,000
> Incendiaries Within 2-Hour Period

May 25, 1945:
(editorial)
> SIX MONTHS OF AIR WAR

The latest attack was in more than five times as great a strength as the first just half a year ago. ...

Their last ally gone, open now to attack by the Allies' full might ... unhappy are the prospects for the little men who thought that they were born to rule the world.

May 26, 1945:
> 500 'SUPERFORTS' FIRE TOKYO'S HEART

May 29, 1945:
> 'SUPERFORTS' DROP FIRE ON YOKOHAMA

May 30, 1945:

TOKYO ERASED, SAYS LEMAY
51 Square Miles Burned Out in Six B-29
Attacks on Tokyo . . .
1,000,000 Japanese Are Believed to Have
Perished in Fires

May 30, 1945:
(editorial)

STRATEGIC BOMBING OF JAPAN

One by one the Twentieth Air Force is ticking off the major Japanese industrial cities. . . . but the campaign is only starting. There is still much to do. . . .

The 20,000-foot-high smoke clouds towering over Yokohama are a signpost on the road to victory, but that road leads far ahead.

August 1, 1945:
NAGOYA AREA HEAVILY BOMBED AGAIN . . .

August 2, 1945:
820 B-29's DROP 6,632 TONS ON FOE . . .
WORLD PEAK BLOW

August 3, 1945:
BOMBERS FIRE GREAT NAGASAKI SHIPYARDS

August 7, 1945:
FIRST ATOMIC BOMB DROPPED ON JAPAN;
MISSILE IS EQUAL TO 20,000 TONS OF TNT;
TRUMAN WARNS FOE OF A 'RAIN OF RUIN' . . .
HIROSHIMA IS TARGET

August 7, 1945:
(editorial)

OUR ANSWER TO JAPAN

The American answer to Japan's contemptuous rejection of the Allied surrender ultimatum of July 26 has now been delivered upon Japanese soil in the shape of a new weapon of destruction which unleashes against it the forces of the universe. . . . What every military man has dreamed of . . . the decisive "secret weapon," the magic key to victory—has been found in America and is now ready to be hurled against our enemies. . . .

So far only one of these bombs has been dropped. . . . That is just a sample of what is in store for all Japan. . . .

It remains to be seen whether the descendants of the Samurai . . . prefer to sacrifice the nation to their fanaticism.

2. Shortly after the end of the war, a survey was taken of the American public's attitudes toward the use of the atomic bomb against Japan. The results, by percentages, were published in *Fortune* magazine in December, 1945:[2]

---

[2] "The Fortune Survey," Courtesy of *Fortune* Magazine, December, 1945.

Which of these comes closest to describing how you feel about our use of the atomic bomb?

1. We should not have used any atomic bombs at all. 4.5
2. We should have dropped one first on some unpopulated region, to show the Japanese its power, and dropped the second one on a city only if they hadn't surrendered after the first one. 13.8
3. We should have used the two bombs on cities, just as we did. 53.5
4. We should have quickly used many more of them before Japan had a chance to surrender. 22.7
5. Don't know. 5.5

3. American public feeling about the Japanese was manifested in the popular songs of the era. One of the best known had the following simple but significant lyrics, sung to a rousing martial tune:[3]

> REMEMBER PEARL HARBOR
> Let's remember Pearl Harbor,
> As we go to meet the foe.
> Let's remember Pearl Harbor,
> As we did the Alamo.
> We will always remember
> How they died for liberty.
> Let's remember Pearl Harbor
> And go on to victory.

4. The author of the book from which the following selection was drawn is a Protestant clergyman in America and an authority in the field of Christian social ethics.[4]

The decision to use the atomic bomb ... took place within a larger context, a significant characteristic of which was the tendency on the part of American leaders (both military and civilian) to think of the war in purely military rather than in political terms. As a result, the goal of the war was military victory; the way to victory was the military defeat of the enemy armed forces; since the major decisions to be made were of a military character, policy formation was delegated to military men. The result was a whole series of decisions: to treat the atomic bomb as a bigger and better military weapon, to exclude civilian leaders from the highest war councils of the nation, to accept the judgment of generals that obliteration bombing was necessary for victory, to try a political method to end the war only as an afterthought, to ignore vital intelligence because it did not indicate that the enemy was ready to accept total surrender, to let a military commander decide on purely tactical grounds how much time the Japanese government should have to decide to surrender after the first atomic bomb—a series of decisions which tended to mold events in such a way that the atomic bomb would finally be used in a total and unrestrained military manner. Any one

---

[3] Lyrics by Charles Newman, Music by Allie Wrubel. Copyright (c) January 30, 1942 by Melrose Music Corp. Used by permission.
[4] Robert C. Batchelder, *The Irreversible Decision 1939-1950* (Boston, Houghton Mifflin, Copyright (c) 1961 by Robert C. Batchelder), pp. 211-215, 217-219.

of these decisions, had the emphasis lain on the political rather than on the military way of thinking, *might* have precluded or tempered our use of the atomic bomb; the cumulative effect of making all these decisions on political rather than on military grounds could well have resulted in termination of the war by diplomatic means, and prevented the destruction of Hiroshima and Nagasaki.

Focusing more narrowly upon the specific decision to drop the atomic bomb—and taking as given the political failure and the war situation as it stood in mid-1945—one cannot escape the conclusion that the atomic bombing of Hiroshima and Nagasaki caused less loss of life (and general human suffering and chaos) than would have come about had the new weapon been withheld and the war been allowed to continue by conventional means, with or without invasion of the Japanese home islands. Within this narrow context Truman and Stimson were right: the atomic bomb did cut short the war and save thousands of lives. Nevertheless, even within the situation as it had developed by August 1945, alternatives were still open —such as a demonstration of the atomic bomb against a large military installation in Japan, followed by a stern warning—which probably would have brought about Japan's surrender without the great toll of civilian lives resulting from atomic attacks upon two large cities.

It was noted . . . [earlier] that certain of the cherished ethical principles held by scientists were transformed under the impact of the threat posed to the civilized world by Hitler's emerging power. Something of the same transformation was wrought in the ethical principles of United States leaders, both military and political, during World War II. Our Air Force had entered the war proud of its Norden bombsight, and was committed to the superior morality and military effectiveness of daylight precision bombing of purely military objectives. Our government was on record as opposed, on humane grounds, to the bombing of civilian areas. The transformation both of practice and of official justification, under the demands of "military necessity," has been documented above. Churchmen proved only slightly more resistant than political leaders to this erosion of moral principles during wartime.

Henry L. Stimson's memoirs and diary record the feelings of one who was sensitive to this transformation and resisted it, yet found himself being swept along and finally participated in it. When the B-29's began their first raids over Japan, Stimson had struck from the list of targets the cultural and religious center of Kyoto, despite the fact that it was also an important center of war industry. He believed that air power could and should be used in a limited and discriminate way—and he extracted from Robert Lovett, his Assistant Secretary for Air, a promise that the B-29's would conduct "only precision bombing" against purely military objectives. His aim, as he put it, was to maintain "the reputation of the United States for fair play and humanitarianism." But then he found that our precisely dropped bombs were falling on soldiers, civilian workers, and children alike, that flames were devouring industry, commercial districts, and private residences indiscriminately. When Stimson called in Arnold to find out "the facts" and to hold him to "my promise from Lovett," the General told him that in the congested Japanese cities it was virtually impossible to destroy war output without also destroying civilians connected with that output. The Secretary could only say that the promise he had extracted should be honored "as far as possible."

Painfully Stimson realized that in the conflagration bombings by massed B-29's he was permitting a kind of total war he had always hated, and in recommending the use of the atomic bomb he was implicitly confessing that there could be no significant limits to the horror of modern war. The decision was not difficult, in 1945, for peace with victory was a prize that outweighed the payment demanded. But Stimson could not dodge the meaning of his action. . . .

Two fundamental ways of judging the morality of the use of the atomic bomb appeared during 1945 and 1946. The first was (in the broadest sense) utilitarian; the primary concern of those using this approach was the consequences of the act in question. Will the war be shortened? How many lives will be lost? Will long-term consequences be good or evil? The method is calculative: good and evil consequences are balanced one against the other, and the right act is that which produces the most good—or, at any rate, the least evil. American's leaders used this method in determining to drop the atomic bomb on Japan. It was the choice, made with awareness of its inherent horror, of the lesser evil. . . .

The second basic ethical approach to the question of the atomic bombing of Japan was formalistic: it was concerned with the rightness or wrongness of the act in itself. What determines the rightness or wrongness of an act is not its consequences but its inherent quality. If the act conforms to an objective moral standard, it is permissible; if not, it is forbidden or condemned. The standard to be applied in the case of Hiroshima may be summed up in the commandment "Thou shalt not attack noncombatants directly." . . .

Despite the immediate clarity of moral judgment provided by the formalistic approach to ethics, it is not devoid of dilemmas. In the example just cited it is too simple, in that it would leave out of consideration the consequences flowing from the two raids. An incendiary raid on Hiroshima, like those on dozens of other Japanese cities during the summer of 1945, would have advanced the end of the war somewhat, but only imperceptibly. In contrast, the atomic raid, as we have seen, was the decisive factor that brought the war to a halt within eight days. Although in one dimension (that of form) the two raids are morally equivalent, in another dimension (that of consequences) one is morally much better than the other. To be concerned only for form and to ignore consequences is to miss much of ethical significance; for certainly it was better on moral grounds that the killing and the disruption of Japanese civil life should stop than that it should continue.

Again, let it be assumed, for the sake of argument, that Truman's estimate of the alternatives before him was accurate: it was a choice between dropping the atomic bomb on Hiroshima and proceeding with the invasion of Japan. A formalist would be compelled to judge the atomic bombing of Hiroshima as impermissible, and therefore to recommend the invasion, which would be justifiable since it would proceed by conventional and discriminate attack on the military forces of the enemy. Having accepted as justifiable the killing of 317,000 Japanese soldiers in the Philippines campaign, and the killing of 107,000 of the total garrison of 120,000 on Okinawa, the Roman Catholic position would now forbid the killing of 110,000 civilians with the atomic bomb but condone a conventional invasion even though the preliminaries and the invasion itself took the lives of ten times that number of Japanese and Allied soldiers in Manchuria, China, and the Japanese home islands. In addition, it would reluctantly "permit" the death of many civilians and the destruction of many cities from battle causes, the freezing and starvation of refugee children during the approaching winter, and the complete breakdown and disruption of the fabric of Japanese civil life—provided only that all these evils were unintended and unavoidable effects of the direct attack of the invaders upon the defenders. . . .

In contrast to a moral theory that would condone the greater evil consequences, provided only that they be produced by "legitimate" means, one cannot help feeling a certain respect for the elemental morality of Truman and Stimson in their determination to avoid the massive evil of invasion if humanly possible. They realistically surveyed the situation confronting them, estimated (probably correctly) that had they refrained from using the atomic bomb the war would have continued without much change in basic character—and then resolutely chose what appeared to them to be the lesser evil, regardless of the fine points of ethical theory. . . .

The conflict between the calculative and formalistic approaches to the bombing of Hiroshima can be focused in a single question: is it right to perform an inherently immoral act in order to achieve a good end and avoid a massive evil? For the formalists the answer is easy—no, it is never permissible to do evil as a means to a good end. Considered in the abstract the problem is simple. But in a particular historical context the answer is not so simple.

The bare act of dropping an atomic bomb upon a city—considered in itself alone—is clearly immoral because it constitutes a direct attack upon noncombatants. Yet it is inconsistent to single out for condemnation the act of dropping an atomic bomb and at the same time implicitly to recommend continuation of a war that one knows will include direct attack upon noncombatants with incendiary bombs. In the midst of a historical context already compromised by past and present mass bombing of civilians, which would undoubtedly have continued in the future, can Truman justly be condemned for authorizing an atomic attack (no more and no less immoral than the fire raids) which promised to put an end to the whole badly compromised situation?

5. On the twentieth anniversary of Hiroshima, in 1965, Pope Paul VI made the statement described in the following article from *The New York Times*: [5]

CASTEL GANDOLFO, Italy, Aug. 8— . . . Speaking from the loggia of his summer residence here, the Pope said, "We pray that that homicidal weapon may not have killed peace in the world; may not have injured forever the honor of science, and that it has not extinguished the serenity of life on this earth."

He prayed that "the world would never see again such an unfortunate day as that of Hiroshima" and that men would "no longer place their trust, their calculations and their prestige in such fatal and dishonoring weapons.". . .

---

[5] *The New York Times*, August 9, 1965, pp. 1-2.

B. THE CONSCIENCE OF SCIENCE

There is irony in the fact that two of the earliest and severest critics of the decision to use the bomb were the two scientists whose letter to President Roosevelt in 1939 started atomic bomb research in this country, Albert Einstein and Leo Szilard. Szilard's views have been encountered previously in this unit. The following documents include a selection from Einstein, followed by dissenting views on the part of two other scientists who contributed importantly to the making of the bomb.

1. Albert Einstein may be said to bear a double responsibility for the atomic bomb. First, it was his epoch-making announcement of the "special theory of relativity" in 1905 that launched the atomic age. He startled the world by proving that matter and energy are equivalent and interchangeable. The atomic bomb may be described as a demonstration of his theory, for its explosion results from a conversion of matter into energy. Secondly, the 1939 letter bore his signature.

After the war Einstein played a leading role in the scientists' movement for nuclear weapons control. In 1946 he gave an interview from which this statement is excerpted:[6]

> The war which began with Germany using weapons of unprecedented frightfulness against women and children ended with the United States using a supreme weapon killing thousands at one blow....
>
> Before the raid on Hiroshima, leading physicists urged the War Department not to use the bomb against defenseless women and children. The war could have been won without it. The decision was made in consideration of possible future loss of American lives—and now we have to consider possible loss in future atomic bombings of *millions of lives.* The American decision may have been a fatal error, for men accustom themselves to thinking a weapon which was used once can be used again.
>
> Had we shown other nations the test explosion at Alamagordo, New Mexico, we could have used it as an education for new ideas. It would have been an impressive and favorable moment to make considered proposals for world order to end war. Our renunciation of this weapon as too terrible to use would have carried great weight in negotiations and made convincing our sincerity in asking other nations for a binding partnership to develop these newly unleashed powers for good....

2. J. Robert Oppenheimer was director of the atomic bomb laboratory at Los Alamos and a member of the Scientific Panel which unanimously recommended use of the bomb without a warning demonstration. He was interviewed in 1965:[7]

> *Q.* After what has happened during these past 20 years, would you, under conditions as they were in 1942, accept once again the invitation to work on the development of the atomic bomb?
> *A.* Yes.
> *Q.* Even after Hiroshima?
> *A.* Yes.

---

[6]Albert Einstein, in an interview with Michael Amrine, "The Real Problem Is in The Hearts of Men," *The New York Times Magazine*, June 23, 1946, pp. 42, 43.
[7]William L. Laurence, "Would You Make The Bomb Again?", *New York Times Magazine*, August 1, 1965, p. 8.

Q. Do you think it was necessary to drop the two atomic bombs over Japan when Japan was already on her knees?

A. From what I know today, I do not believe that we could have known with any degree of certainty that the atomic bomb was necessary to end the war. But that was not the view of those who had studied the situation at the time and who were thinking of an invasion of Japan. Probably they were wrong. Japan had already approached Moscow to sound out the United States about terms of peace. Probably a settlement could have been reached by political means. But the men who made the decision—and I am thinking particularly of Secretary Stimson and President Truman—were sure that the choice was either invasion or the bomb. Maybe they were wrong, but I am not sure that Japan was ready to surrender.

Dr. Oppenheimer added:

"I never regretted, and do not regret now, having done my part of the job. I have a deep, continuing, haunting sense of the damage done to European culture by the two world wars. The existence of the bomb has reduced the chance of World War III and has given us valid hope.

"I believe it was an error that Truman did not ask Stalin to carry on further talks with Japan, and also that the warning to Japan was completely inadequate.

"But I also think that it was a damn good thing that the bomb was developed, that it was recognized as something important and new, and that it would have an effect on the course of history. In that world, in that war, it was the only thing to do. I only regret that it was not done two years earlier. It would have saved a million or more lives."

3. Luis W. Alvarez, a theoretical physicist who had done brilliant work at the University of California, played a major role in the Los Alamos laboratory. He also helped assemble the two atomic bombs for the flights to Japan and flew aboard the B-29 *Enola Gay* on its bomb-run over Hiroshima. In 1968 he won the Nobel Prize in physics. Following are excerpts from an interview that took place in 1965:[8]

Q. Did you approve the use of the bomb in 1945?

A. Of course. We had been in the war a long time. It seemed certain to continue for a long time, with enormous loss of life on both sides. We had the means to end the war quickly, with a great saving of human life. I believed it was the only sensible thing to do, and I still do.

Q. Would you do it over again? Would you still work on the bomb?

A. Of course. The weapon was possible. So far as we knew, we were in a race with the Germans. We had to beat them to it, or risk losing the war.

Q. Would you have done it knowing subsequent history?

A. Yes. I am proud to have had a part in a program that by most modern estimates saved a million lives, both Japanese and American, that would otherwise have been lost in the projected invasion. This pride is reinforced by the knowledge that the world has not had a major war in the past 20 years, and that most responsible people feel the risk of a World War III has diminished steadily with time in these same years. I am confident that both of these admirable situations are directly traceable to the existence of nuclear weapons.

---
[8] *Ibid.,* p. 53.

C. THE CONSCIENCE OF SOLDIERS

In this final group of documents, military men of widely varying rank and from several branches of the armed forces express their personal feelings as to the rights and wrongs of the bombing of Hiroshima.

1. General Eisenhower in his book *Crusade in Europe*, published in 1948, describes how he felt about the atomic bomb shortly before Hiroshima:[9]

> I had a long talk with Secretary Stimson, who told me that very shortly there would be a test in New Mexico of the atomic bomb, which American scientists had finally succeeded in developing. The results of the successful test were soon communicated to the Secretary by cable. He was tremendously relieved, for he had apparently followed the development with intense interest and felt a keen sense of responsibility for the amount of money and resources that had been devoted to it. I expressed the hope that we would never have to use such a thing against any enemy because I disliked seeing the United States take the lead in introducing into war something as horrible and destructive as this new weapon was described to be.

2. Lieutenant General Leslie R. Groves directed the entire 2 billion dollar Manhattan Project which produced the atomic bomb. These excerpts are from a 1965 interview:[10]

> There was never any question as to the use of the bomb, if it was successfully developed, on the part of anyone who was in a top position on the project and who knew what was going on....
>
> Above all else was the very strong feeling on the part of President Truman, which was the same feeling that the rest of us who knew about it had, that it was criminal and morally wrong for us to have means to bring this war to a proper conclusion and then not use the means.
>
> It is true we didn't need to bomb to win, but we needed it to save American lives.

3. In an earlier section of this unit, Admiral Leahy, Chief of Staff to Presidents Roosevelt and Truman, expressed his opposition to use of the bomb on military grounds. Here he registers a vigorous protest to its use on moral grounds:[11]

> "Bomb" is the wrong word to use for this new weapon. It is not a bomb. It is not an explosive. It is a poisonous thing that kills people by its deadly radioactive reaction, more than by the explosive force it develops.
>
> The lethal possibilities of atomic warfare in the future are frightening. My own feeling was that in being the first to use it, we had adopted an ethical standard common to the barbarians of the Dark Ages. I was not taught to make war in that fashion, and wars cannot be won by destroying women and children....
>
> These new concepts of "total war" are basically distasteful to the soldier and sailor of my generation. Employment of the atomic bomb in war will take us back in cruelty toward noncombatants to the days of Genghis Khan.

---
[9] Dwight D. Eisenhower, *Crusade in Europe*, p. 443.
[10] William L. Laurence, "Would You Make The Bomb Again?", p. 9.
[11] William D. Leahy, *I Was There*, pp. 441-442.

It will be a form of pillage and rape of a society, done impersonally by one state against another, whereas in the Dark Ages it was a result of individual greed and vandalism. These new and terrible instruments of uncivilized warfare represent a modern type of barbarism not worthy of Christian man.

One of the professors associated with the Manhattan Project told me that he had hoped the bomb wouldn't work. I wish that he had been right.

4. Less than a month before Hiroshima, the Japanese announced the formation of a People's Volunteer Corps, making all men from 15 to 60 and women from 17 to 40 liable for defense duties. The following statement, which appeared in an official Army Air Force publication about two weeks before the atomic bomb was dropped, showed one American's reaction to this news, probably typifying the feelings of many men in the armed services and even of the general public:[12]

> The entire population of Japan is a proper Military Target . . ., THERE ARE NO CIVILIANS IN JAPAN. We are making War and making it in the all-out fashion which saves American lives, shortens the agony which War is and seeks to bring about an enduring Peace. We intend to seek out and destroy the enemy wherever he or she is, in the greatest possible numbers, in the shortest possible time.

5. Abe Spitzer was a B-29 radio operator who flew both the Hiroshima and Nagasaki missions. In a book published the following year, Spitzer recorded his own and his buddies' reactions:[13]

> "The part that got me most," Ray commented, "was when I saw that ball of fire. I guess I can tell you now, and you won't laugh, but I thought maybe the world had come to an end, and we'd caused it."
>
> "I can't forget that pillar of smoke," said Buckley, "reaching all the way up and not getting any place, not ever getting to the top of the sky, and I kept wondering if it would ever stop. And I wouldn't want to be too sure if it ever did."
>
> "Funny," added Pappy, "I couldn't help thinking about the people who'd been living in those houses down there. They never knew what hit them. I knew they were Japs, of course, and you don't think much about Japs. They're your enemies and all. But I kept thinking about that, and I still do. I guess I always will."
>
> "I just thought about maybe I should have stayed in the tank outfit," said Kuharek. "That's a nice, clean way to kill people. If it's all right to kill people, that's a nice, clean way to do it. I mean then they can fight back."
>
> I didn't say much . . . but now, thinking over the conversation of that evening, it seems to me Ray said it best.
>
> ". . . I thought maybe the world had come to an end, and we'd caused it."
>
> Well, a world did come to an end, although we hadn't caused it. We were under orders; we were just doing our job. And I don't think Ray meant it

---

[12]Col. Harry F. Cunningham, A-2 of the Fifth Air Force (Fifth Air Force Weekly Intelligence Review, No. 86, 15-21 July 1945), cited in Craven and Cate, *The Army Air Forces in World War II,* V, pp. 696-697n.
[13]Merle Miller and Abe Spitzer, *We Dropped the A-Bomb* (Thomas Y. Crowell Company, Copyright 1946 by Merle Miller and Abe Spitzer), pp. 151-152.

in exactly the way I do, but he was close enough. And that's what I've been trying to tell people ever since, not very successfully; some people don't listen, and some people don't care, and hardly anyone, including me, knows exactly what to do about the fact. All I know is that it's about time we found out.

And I also know that I agree with the scientist back at Tinian who, when asked how he felt about his part in making the bomb, said, "I'm not proud of myself right now." Neither am I.

The missions to Hiroshima and Nagasaki are nothing to be proud of at all. But they are missions to be remembered. And never repeated. Anywhere in the world. Ever again.

SUGGESTIONS FOR FURTHER READING

The best-known account of how Hiroshima's people were affected by the bomb is John Hersey's *Hiroshima* (New York, Knopf, 1946). A Japanese doctor's experiences are related by M. Hachiya in *Hiroshima Diary* (Chapel Hill, University of North Carolina Press, 1955). In *Nuclear Disaster* (New York, Meridian Books, 1963), Tom Stonier describes the probable effects of a thermonuclear bomb on an American city.

Opposing views on the military controversy can be found in H. W. Baldwin, *Great Mistakes of the War* (New York, Harper, 1950) and in an article by Louis Morton in Kent Roberts Greenfield, ed., *Command Decisions* (New York, Harcourt, Brace and World, 1959). The man who headed the Manhattan Project, General Leslie R. Groves, vigorously defends the use of the bomb in *Now It Can Be Told* (New York, Harper, 1962). A more personal account is *Atoms in the Family* (Chicago, University of Chicago Press, 1961) by Laura Fermi, widow of Enrico Fermi. A brilliant study of two leading atomic scientists is N. P. Davis, *Lawrence and Oppenheimer* (New York, Simon and Schuster, 1968). The troubled relations between scientists and government are analyzed in the official history of the Atomic Energy Commission, R. G. Hewlett and O. G. Anderson, *The New World, 1939-1946* (University Park, Pennsylvania State University Press, 1962).

Two incisive studies of the diplomatic problem are R. J. C. Butow, *Japan's Decision to Surrender* (Stanford, Stanford University Press, 1954 and Herbert Feis, *Japan Subdued* (Princeton, Princeton University Press, 1961). The last-ditch attempt by Japanese army officers to prevent the surrender through a coup d'etat is described by William Craig in *The Fall of Japan* (New York, Dial Press, 1967). An extensive analysis of the Russian problem, critical of American foreign policy including our use of the bomb, is D. F. Fleming, *The Cold War and Its Origins* (London, Doubleday, 1961). Defenses of U. S. policy are too numerous to list here.

The *Bulletin of the Atomic Scientists* is a consistently provocative periodical in this field. In *Look*, August 13, 1963, F. Knebel and C. Bailey argue that the scientists' protests were prevented from reaching the President. The entire August 1, 1965 issue of *The New York Times Magazine* is devoted to the 20th anniversary of Hiroshima.

Collections of readings include E. Fogelman, *Hiroshima: The Decision to Use the A-bomb* (New York, Scribner, 1964) and P. R. Baker, *The Atomic Bomb* (New York, Holt, Rinehart and Winston, 1968).

The National Broadcasting Company has produced two excellent television documentaries, "The Decision to Drop the Bomb" and "The Surrender of Japan." The research materials for these programs form the basis of the book by L. Giovanitti and F. Freed, *The Decision to Drop the Bomb* (New York, Coward-McCann, 1965), which surveys the entire problem in readable fashion.

CALDWELL COLLEGE LIBRARY
CALDWELL, NEW JERSEY

ADDISON-WESLEY PUBLISHING COMPANY